Lecture Notes in Statistics

Edited by D. Brillinger, S. Fienberg, J. Gani,
J. Hartigan, J. Kiefer, and K. Krickeberg

6

Shanti S. Gupta and Deng-Yuan Huang

Multiple Statistical Decision Theory: Recent Developments

Springer-Verlag
New York   Heidelberg   Berlin

Shanti S. Gupta
Department of Statistics
Purdue University
West Lafayette, Indiana 47907
U.S.A.

Deng-Yuan Huang
Institute of Mathematics
Academia Sinica, Taipei
Taiwan, Republic of China

AMS Subject Classification: 62C25

Library of Congress Cataloging in Publication Data

Gupta, Shanti Swarup, 1925-
  Multiple statistical decision theory.

  (Lecture notes in statistics; 6)
  1. Sequential analysis.   2. Ranking and
selection (Statistics)   3. Multiple comparisons
(Statistics)   I. Huang, Deng-Yuan.   II. Title.
III. Series:
QA279.7.G87   519.5'4   81-886
                                   AACR2

All rights reserved.

No part of this book may be translated or reproduced in any
form without written permission from Springer-Verlag.

The use of general descriptive names, trade names, trademarks,
etc. in this publication, even if the former are not especially
identified, is not to be taken as a sign that such names, as
understood by the Trade Marks and Merchandise Marks Act, may
accordingly be used freely by anyone.

© 1981 by Springer-Verlag New York Inc.

Printed in the United States of America

9 8 7 6 5 4 3 2 1

ISBN 0-387-90572-3 Springer-Verlag   New York   Heidelberg   Berlin
ISBN 3-540-90572-3 Springer-Verlag   Berlin   Heidelberg   New York

Preface

The theory and practice of decision making involves infinite or finite number of actions. The decision rules with a finite number of elements in the action space are the so-called multiple decision procedures. Several approaches to problems of multiple decisions have been developed; in particular, the last decade has witnessed a phenomenal growth of this field. An important aspect of the recent contributions is the attempt by several authors to formalize these problems more in the framework of general decision theory. In this work, we have applied general decision theory to develop some modified principles which are reasonable for problems in this field. Our comments and contributions have been written in a positive spirt and, hopefully, these will an impact on the future direction of research in this field. Using the various viewpoints and frameworks, we have emphasized recent developments in the theory of selection and ranking which, in our opinion, provides one of the main tools in this field. The growth of the theory of selection and ranking has kept apace with great vigor as is evidenced by the publication of two recent books, one by Gibbons, Olkin and Sobel (1977), and the other by Gupta and Panchapakesan (1979). An earlier monograph by Bechhofer, Kiefer and Sobel (1968) had also provided some very interesting work in this field.

A related topic of interest is multiple comparison theory which has been developed well and can be found in many works. Recent developments in this theory based on stagewise tests have been studied briefly in this monograph.

We wish to thank Professor J. Kiefer for his valuable and constructive suggestions and Professor S. Panchapakesan for helpful discussions. Professor K. J. Miescke has also given some very helpful suggestions and comments which we wish to acknowledge thankfully. Mrs. Norma Lucas has typed and retyped the manuscript several times with great patience and accuracy. For this excellent work, we are, indeed, very grateful to her. Financial assistance for this work has been extended by the Office of Naval Research Contract N00014-75-C-0455 at Purdue University and the authors acknowledge this gratefully. The second author has received partial support and encouragement from Academia Sinica, Republic of China, during the writing of this monograph and would like to express his appreciation to them.

Purdue University	Shanti S. Gupta
August, 1980	and
	Deng-Yuan Huang

Table of Contents

Chapter 1   1

    Some Auxiliary Results: Monotonicity Properties of Probability Distributions

    1.1. Introduction   1
    1.2. Ordered Families of Distributions   1
    1.3. Probability Integrals of Multivariate Normal Distribution - Dependence on Correlations   6
    1.4. Dependence and Association of Random Variables   15
    1.5. Majorization in Multivariate Distribution   21

Chapter 2   29

    Multiple Decision Theory: A General Approach

Chapter 3   38

    Modified Minimax Decision Procedures

    3.1. Introduction   38
    3.2. The Problem of Selecting Good Populations with Respect to A Control   38
    3.3. On the Problem of Selecting the Best Population   43
    3.4. Essentially Complete Classes of Decision Procedures   50

Chapter 4   61

    Invariant Decision Procedures

    4.1. Introduction   61
    4.2. Selecting the Best Population   61

Chapter 5   73

    Robust Selection Procedures: Most Economical Multiple Decision Rules

    5.1. Introduction   73
    5.2. Robust Selection Rules   73

Chapter 6   80

    Multiple Decision Procedures Based on Tests

    6.1. Introduction   80
    6.2. Conditional Confidence Approach   80
    6.3. Multiple Comparison Procedures   84
    6.4. Multiple Range Tests   89
    6.5. Multistage Comparison Procedures   93

Introduction

In statistical inference, multiple decision theory is becoming a very active area of research. It includes very wide subjects such as classification, ranking and selection, identification, multiple comparison and so on. In practice, one often encounters the situation with a finite number of actions, this being the case under study. Significant work has been done under certainty for multi-objective problems. Various applications are well known, especially in economics, management and so on. But it is more practical to consider uncertainty than certainty in most real situations. Studying the case where uncertainty is present is the purpose of the present monograph.

The traditional approach in this area leaves many things open to criticism. We point out some of these and give some reasonable ways to derive "good" procedures. In this monograph, we emphasize the principles of, and the criteria for, multiple decision procedures. Some fundamental known results have been proved and incorporated for convenience. For further details, one should refer to the original papers.

We introduce the concepts of monotonicity of probability distributions and study its consequences and special cases in Chapter 1. These results provide important tools to evaluate the probability integrals and their bounds for multivariate distributions. In Chapter 2, we describe a general approach to multiple decision problems. In Chapter 3, we discuss modifications of the minimax criteria. The reason for the modification is that the minimax risk over all rules may not be sufficient (and efficient) enough to suit the experimenter's objective(s). Thus we study rules restricting our attention to a class of decision rules based on various realistic criteria and goals. Invariant decision rules are considered in Chapter 4. The problem of designing an experiment for determining the minimum sample sizes in the sense of most economical rules is discussed in Chapter 5. In Chapter 6, we deal with multiple decision rules based on statistical tests of hypotheses. A brief description of Kiefer's conditional confidence approach to testing is provided. Some multiple comparison procedures based on stagewise tests are also described in this chapter.

CHAPTER 1
SOME AUXILIARY RESULTS - MONOTONICITY PROPERTIES OF PROBABILITY DISTRIBUTIONS

1.1. Introduction

It is very important to study the monotonicity properties of distributions in order to obtain inequalities useful in statistical inference. Some monotonicity properties of distributions are well known and have proved to be very useful. During the last decade, more concepts have been introduced and used by several authors in multiple decision problems.

In Section 1.2., we discuss the distributions which have stochastically increasing property in a very general setting. The monotonicity of probability integrals of multivariate normal distribution is of special importance for the evaluation of multivariate normal probability integrals. We discuss it in Section 1.3. In Section 1.4, we study various types of monotonicity which are applicable without the assumption of normality. The relations between these various types are discussed. Some examples are given to show that none of the proposed definitions are equivalent and moreover to show the importance of all these concepts. The relation with some older and well-known concepts will also be considered. In Section 5, we discuss the monotonicity of distributions in terms Schur functions. Various useful densities will be given to show the importance of this idea. Applications to inequalities are given by Marshall and Olkin (1979) and Tong (1980).

1.2. Ordered Families of Distributions

Let $\underline{X} = (X_1,\ldots,X_n)$ be a random vector with a probability distribution P_θ, depending on a real parameter θ. In most problems that one encounters in applications, such distributions are usually ordered, roughly speaking, in the sense that large values of θ lead, on the whole, to large values of the X's. The concept of monotone likelihood ratio (MLR) due to Karlin and Rubin (1956) is very important in statistics. The concept of total positivity (see Karlin (1968)) is more general. In the case of total positivity of order 2 (TP_2), if densities exist, then TP_2 is equivalent to MLR. Further, MLR implies stochastically increasing property (SIP).

We now discuss several results relating to the stochastic ordering of distributions and the monotonicity of certain probability integrals.

1.2.1 Definition. A function φ defined on $\mathbb{R}^m$, an m-dimensional Euclidean space, is said to be increasing for a partial order "$\leq$", if $\underline{x}_1 \leq \underline{x}_2$ implies $\varphi(\underline{x}_1) \leq \varphi(\underline{x}_2)$.

1.2.2 Definition. A set S in $\mathbb{R}^m$ is said to be increasing if its indicator set function is increasing, i.e., if $\underline{x}_1 \in S$ and $\underline{x}_1 \leq \underline{x}_2$ then $\underline{x}_2 \in S$.

Lehmann (1955) has shown that the following two conditions are equivalent:
(A) P_θ has SIP,

(B) If $\varphi(\underline{x})$ is an increasing function, then $E_{\underline{\theta}}\varphi(\underline{X})$ is increasing in θ.

Gupta and Panchapakesan (1972) give a more general result to provide sufficient conditions for the monotonicity problems as follows:

1.2.3 <u>Theorem</u>. Let $\{F(\cdot|\lambda), \lambda \in \Lambda\}$ be a family of absolutely continuous distributions on the real line with continuous densities $f(\cdot|\lambda)$ and $\varphi(x,\lambda)$ a bounded real-valued function possessing first partial derivatives ϕ_x and ϕ_λ with respect to x and λ respectively and satisfying regularity conditions (*). Then $E_\lambda[\phi(X,\lambda)]$ is nondecreasing in λ provided for all $\lambda \in \Lambda$,

$$(1.2.1) \quad f(x|\lambda) \frac{\partial \phi(x,\lambda)}{\partial \lambda} - \frac{\partial F(x,\lambda)}{\partial \lambda} \frac{\partial \phi(x,\lambda)}{\partial x} \geq 0 \quad \text{a.e. } x,$$

where

(*) (i) for all $\lambda \in \Lambda$, $\frac{\partial \phi(x,\lambda)}{\partial x}$ is Lebesgue integrable on $\mathbb{R}'$; and

(ii) for every $[\lambda_1,\lambda_2] \subset \Lambda$ and $\lambda_3 \in \Lambda$ there exists $h(x)$ depending only on λ_i, $i = 1,2,3$ such that

$$\left| \frac{\partial \phi(x,\lambda)}{\partial \lambda} f(x|\lambda_3) - \frac{\partial F(x|\lambda)}{\partial \lambda} \frac{\partial \phi(x,\lambda_3)}{\partial x} \right| \leq h(x)$$

for all $\lambda \in [\lambda_1,\lambda_2]$ and $h(x)$ is Lebesgue integrable on $\mathbb{R}'$.

<u>Proof</u>. Let us consider $\lambda_1, \lambda_2 \in \Lambda$ such that $\lambda_1 \leq \lambda_2$ and define

$$(1.2.2) \quad A_i(\lambda_1,\lambda_2) = \int \prod_{\substack{r=1 \\ r \neq i}}^{2} \phi(x,\lambda_r) dF_i(x), \quad i = 1,2$$

and

$$(1.2.3) \quad B(\lambda_1,\lambda_2) = \sum_{i=1}^{2} A_i(\lambda_1,\lambda_2),$$

where $F_i \equiv F_{\lambda_i}$, $i = 1,2$. We note that when $\lambda_1 = \lambda_2 = \lambda$, $B(\lambda,\lambda) = 2E_\lambda \phi(X,\lambda)$.

Integrating $A_1(\lambda_1,\lambda_2)$ by parts and using it in (1.2.3), it is easily seen that

$$(1.2.4) \quad B(\lambda_1,\lambda_2) = \text{a term independent of } \lambda_1$$
$$+ \int \{\phi(x,\lambda_1) f_2(x) - F_1(x) \phi_x(x,\lambda_2)\} dx.$$

Hence,

$$(1.2.5) \quad \frac{\partial}{\partial \lambda_1} B(\lambda_1,\lambda_2) = \int \{\phi_{\lambda_1}(x,\lambda_1) f_2(x) - \frac{\partial}{\partial \lambda_1} F_1(x) \phi_x(x,\lambda_2)\} dx$$

and this is nonnegative if, for $\lambda_1 \leq \lambda_2$,

$$(1.2.6) \quad \phi_{\lambda_1}(x,\lambda_1) f_{\lambda_2}(x) - \frac{\partial}{\partial \lambda_1} F_{\lambda_1}(x) \phi_x(x,\lambda_2) \geq 0$$

for all x.

Now, we consider the configuration $\lambda_1 = \lambda_2 = \lambda$. It can be easily verified that

(1.2.7) $$\frac{d}{d\lambda} B(\lambda,\lambda) = \sum_{i=1}^{2} \frac{\partial}{\partial \lambda_i} B(\lambda_1,\lambda_2)\Big|_{\lambda_1=\lambda_2=\lambda}$$

and

(1.2.8) $$\frac{\partial}{\partial \lambda_2} B(\lambda_1,\lambda_2) = \frac{\partial}{\partial \lambda_2} B(\lambda_2,\lambda_1) = \frac{\partial}{\partial \lambda_1} B(\lambda_1,\lambda_2)\Big|_{\lambda_1 \leftrightarrow \lambda_2},$$

where $\lambda_1 \leftrightarrow \lambda_2$ indicates that after differentiation λ_1 and λ_2 are interchanged in the final expression. Thus

(1.2.9) $$\frac{d}{d\lambda} B(\lambda,\lambda) = 2 \frac{\partial}{\partial \lambda_1} B(\lambda_1,\lambda_2)\Big|_{\lambda_1=\lambda_2=\lambda}.$$

Hence $B(\lambda,\lambda)$ is nondecreasing if (1.2.6) holds. For the theorem, it is easy to see that it suffices if (1.2.6) holds when $\lambda_1 = \lambda_2 = \lambda$ in a manner consistent with (1.2.6); in other words, if (1.2.1) holds. The strict inequality part is now obvious.

Especially, if $\phi(\lambda,x) = \phi(x)$ for all $\lambda \in \Lambda$, then (1.2.1) reduces to

(1.2.10) $$\frac{\partial}{\partial \lambda} F_\lambda(x) \frac{d}{dx} \phi(x) \leq 0.$$

This is satisfied if $\{F_\lambda\}$ is stochastically non-decreasing family of distributions and $\phi(x)$ is nondecreasing and differentiable at x and hence $E_\lambda \phi(X)$ is nondecreasing in λ, which is a result of Lehmann (1959, p. 112).

Now we discuss the case where the parameter space has dimensions k ($\geq$ 1). Here we shall consider monotonicity properties for multivariate distributions as follows: Let $P_{\underline{\theta}}(S)$ denote, for convenience, $P_{\underline{\theta}}(\underline{X} \in S)$ for a measurable set S where $\underline{\theta} = (\theta_1,\ldots,\theta_k)$.

1.2.4 __Definition__. A distribution function is said to have stochastically increasing property (SIP) in $\underline{\theta}$ if

$$P_{\underline{\theta}_1}(S) \leq P_{\underline{\theta}_2}(S)$$

for every monotone measurable nondecreasing set S and for every $\underline{\theta}_1 \leq \underline{\theta}_2$.

Note that in case of independence it is well known that MLR implies SIP (see Lehmann (1959), p. 74). This can be shown to be true also for several cases of generalized MLR when there is no independence.

As before let E denote the expectation with respect to the distribution $P_{\underline{\theta}}$. It is easy to show the following result (cf. Lehmann (1959) and Hoel (1970)).

1.2.5 __Theorem__. Let the distribution of $\underline{X}$ have stochastically increasing property in $\underline{\theta}$, and let $\varphi(\underline{x},\underline{\theta})$ be non-decreasing in $\underline{x}$ and $\underline{\theta}$. Then $E_{\underline{\theta}}\varphi(\underline{X},\underline{\theta})$ is non-decreasing in $\underline{\theta}$.

Now, we discuss another monotonicity property of multivariate density functions. Consider a Borel subset $\chi \subset \mathbb{R}^k$ and let μ be a σ-finite measure on χ. Also, let Θ be a symmetric subset of $\mathbb{R}^k$ and G be an arbitrary set. For a family of density functions (wrt μ) $\{f_{\underline{\alpha}}(\underline{x},\underline{\theta}) | \underline{\alpha} \in G\}$ where $\underline{\alpha}$ is a nuisance parameter vector, Eaton (1967) introduces the following concept which is also of use in the treatment of ranking problems.

1.2.6 **Definition**. A family of real-valued density functions $\{f_\alpha(\underline{x},\underline{\theta})\}$ is said to have property M if for each $\underline{\alpha} \in G$ and for each i,j ($i \neq j$), $1 \leq i$, $j \leq k$, the following holds:

(1.2.11)
$$x_i \geq x_j \text{ and } \theta_i \geq \theta_j \text{ imply that}$$
$$f_\alpha(\underline{x},\underline{\theta}) \geq f_\alpha(\underline{x}, (i,j)\underline{\theta})$$

where $(i,j)\underline{\theta}$ is the vector with the components θ_i and θ_j interchanged, other components being fixed.

It is easy to see that the family $\{f_\alpha | \underline{\alpha} \in G\}$ with

(1.2.12)
$$f_\alpha(x_1,\ldots,x_k; \theta_1,\ldots,\theta_k) = \prod_{i=1}^{k} g_\alpha(x_i,\theta_i),$$

has property M if and only if $\{g_\alpha | \underline{\alpha} \in G\}$ has MLR.

The above result (1.2.12) is helpful in the construction of many densities $\{f_\alpha\}$ which have property M. However, there are multivariate densities of interest which, of course, cannot be written in the form (1.2.12) because of the lack of independence. Eaton (1967) gives a necessary and sufficient condition for a class of densities to possess the property M.

1.2.7 **Theorem**. Let f be a positive strictly decreasing function defined on $[0,\infty)$ and consider

(1.2.13)
$$h_\Lambda(\underline{x},\underline{\theta}) = c(\Lambda)f((\underline{x}-\underline{\theta})'\Lambda(\underline{x}-\underline{\theta}))$$

where $\underline{\theta}$ and $\underline{x}$ are (column) vectors in $\mathbb{R}^k$, ($k \geq 2$), Λ is a k×k positive definite matrix and $c(\Lambda)$ is a positive constant. Assume that $h_\Lambda(\underline{x},\underline{\theta})$ is a density on $\mathbb{R}^k$. Such a density is also called an elliptically contoured density. Then, a necessary and sufficient condition for the density $h_\Lambda(\underline{x},\underline{\theta})$ to have the property M is that $\Lambda = c_1 I - c_2 ee'$, where $e' = (1,1,\ldots,1)$, $c_1 > 0$, $-\infty < c_2/c_1 < 1/k$, and I is the k×k identity matrix.

Proof. Assume that the condition holds. The conditions are equivalent on c_1 and c_2 are equivalent to the positive definiteness of Λ. For $x_i \geq x_j$ and $\theta_i \geq \theta_j$ ($i \neq j$), a straightforward computation yields
$$(\underline{x}-\underline{\theta})'\Lambda(\underline{x}-\underline{\theta}) \leq (\underline{x}-(i,j)\underline{\theta})'\Lambda(\underline{x}-(i,j)\underline{\theta})$$
when $\Lambda = c_1 I - c_2 ee'$. Since f is a decreasing function.

(1.2.14)
$$h_\Lambda(\underline{x},\underline{\theta}) \geq h_\Lambda(\underline{x}, (i,j)\underline{\theta}),$$

for $x_i \geq x_j$ and $\theta_i \geq \theta_j$. Note that in the above we have only used that f is decreasing (not that f is strictly decreasing).

Conversely, suppose h_Λ has property M. For each i,j ($i \neq j$), $x_i \geq x_j$ and $\theta_i \geq \theta_j$ implies (1.2.14), or equivalently,
$$-2\underline{x}'\Lambda\underline{\theta} + \underline{\theta}'\Lambda\underline{\theta} \leq -2\underline{x}'\Lambda[(i,j)\underline{\theta}] + [(i,j)\underline{\theta}]'\Lambda[(i,j)\underline{\theta}].$$

Setting $\underline{x} = \underline{0}$, $\underline{\theta}' = (1,0,\ldots,0)$, $(i,j) = (1,2)$ yields $\lambda_{11} \geq \lambda_{22}$; while $\underline{x} = \underline{0}$, $\underline{\theta}' = (0,1,0,\ldots,0)$, $(i,j) = (2,1)$ yields $\lambda_{22} \geq \lambda_{11}$ so that $\lambda_{11} = \lambda_{22}$. Similarly, all the diagonal elements of Λ are equal, say $\lambda_{ii} = \lambda$, $i = 1,2,\ldots,k$.

Now, setting $\underline{x} = \underline{0}$, $\underline{\theta}' = (1,1,0,\ldots,0)$, $(i,j) = (1,3)$ yields $2(\lambda+\lambda_{12}) \leq 2(\lambda+\lambda_{23})$; setting $\underline{x} = \underline{0}$, $\underline{\theta}' = (0,1,1,0,\ldots,0)$, $(i,j) = (3,1)$ yields $2(\lambda+\lambda_{23}) \leq 2(\lambda+\lambda_{12})$ so that $\lambda_{12} = \lambda_{23}$. In a similar manner, all the off diagonal elements of Λ are shown to be equal. The proof is completed by noting that Λ is assumed to be positive definite.

It should be pointed out that if θ is a location parameter in the distribution of $\underline{X}$ then the distribution of $\underline{X}$ has SIP but not necessarily the property M. On the other hand, the multinomial distribution has both SIP and the property M (see Alam (1973)). The function defined as in (1.2.13) is Schur-concave which is defined in Section 1.5.

Hollander, Proschan and Sethuraman (1977) have defined a concept of decreasing in transposition (DT) which is the same as the property M. They prove the following result: Let $g(\underline{\lambda},\underline{x}) = h(\underline{\lambda}-\underline{x})$. Then g is DT on $\mathbb{R}^{2n}$ if and only if h is Schur-concave on $\mathbb{R}^n$.

The following example due to Hsu (1977) shows that the property M does not imply SIP.

1.2.8 Example. Let $\underline{X} = (X_1, X_2)$ and $\underline{\theta} = (\theta_1, \theta_2)$.

$\underline{\theta}$ \ $\underline{x}$	(5,6)	(6,5)
(1,2)	0.9	0.1
(2,1)	0.1	0.9
(3,4)	0.6	0.4
(4,3)	0.4	0.6

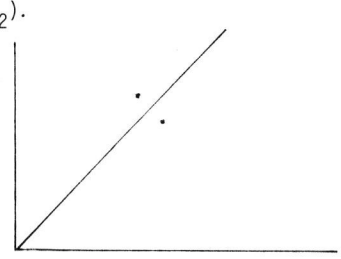

This family of four distributions clearly has property M. But it does not have SIP. For instance,

$$P\{X_1 \geq 5, X_2 \geq 6 | (1,2)\} = P\{(5,6)|(1,2)\} = 0.9$$
$$> 0.6 = P\{(5,6)|(3,4)\} = P\{X_1 \geq 5, X_2 \geq 6|(3,4)\}.$$

Similar to the idea of MLR property for the case of univariate density, Oosterhoff (1969) defines a strict monotone likelihood ratio density and Gupta and Huang (1980) define a generalized monotone likelihood ratio property which will be used latter. Also, some (related) concepts of total monotone likelihood ratio (TMLR) and total stochastic monotone property (TSMP) have been defined by Gupta and Hsu (1977).

1.3. Probability Integrals of Multivariate Normal Distributions - Dependence on Correlations

We shall now study the monotonicity of distributions depending on correlations which is a special form of $f_\theta(\underline{x})$ as described in Section 1.2. Some work has been done on the evaluation of multivariate normal integrals in higher dimensions over domains of the type
$$[-\infty < x_1 \leq h_1, -\infty < x_2 \leq h_2, \ldots, -\infty < x_n \leq h_n].$$
As n, the number of variables, increases, the number of correlations, which is equal to $\binom{n}{2}$, increases rapidly. It is difficult to compute the integrals described above except when the correlation matrix has some kind of special pattern. We are interested in discussing some monotonicity properties of the probability integrals and get a bound under some special forms of correlations.

Let $(X_1, X_2, \ldots, X_n)$ be a random vector having n-variate normal distribution with mean values 0, variances 1. Let

(1.3.1)
$$\Phi_n(h_1, \ldots, h_n; \Sigma)$$
$$= \int_{-\infty}^{h_1} \cdots \int_{-\infty}^{h_n} g_n(x_1, \ldots, x_n; \Sigma) dx_1, \ldots, dx_n,$$

where
$$g_n(\underline{x}; \Sigma) = (2\pi)^{-\frac{n}{2}} |\Sigma|^{-\frac{1}{2}} \exp\{-\frac{1}{2} \underline{x}' \Sigma^{-1} \underline{x}\},$$
and $\underline{x}' = (x_1, \ldots, x_n)$, $\Sigma = \{\rho_{ij}\}$ is the positive definite correlation matrix of the X_i's.

A monotonicity property concerning $\Phi_n(a_1, a_2, \ldots, a_n; \Sigma)$ has been discussed by Gupta (1963) as follows.

1.3.1 __Theorem__ (Slepian (1962)). Let $(X_1, X_2, \ldots, X_n)$ be multivariate normal with zero means and positive definite covariance matrix $\Sigma_1 = \{\rho_{ij}\}$ and let $Y_1, \ldots, Y_n$ be multivariate normal with zero means and positive definite covariance matrix $\Sigma_2 = \{\kappa_{ij}\}$. Let $\rho_{ii} = \kappa_{ii} = 1$, $i = 1, 2, \ldots, n$. If $\rho_{ij} \geq \kappa_{ij}$, $i,j = 1, 2, \ldots, n$, then
$$P\{X_1 > a_1, X_2 > a_2, \ldots, X_n > a_n\}$$
$$\geq P\{Y_1 > a_1, Y_2 > a_2, \ldots, Y_n > a_n\},$$
and

(1.3.2)
$$\Phi_n(a_1, a_2, \ldots, a_n; \Sigma_1) \geq \Phi_n(a_1, a_2, \ldots, a_n; \Sigma_2).$$

__Proof.__ By writing the density function g_n as the inverse of its characteristic function, it is easy to verify that if Σ_1 is positive definite,

(1.3.3) $$\frac{\partial g_n}{\partial \rho_{ij}} = \frac{\partial^2 g_n}{\partial x_i \partial x_j},$$

(1.3.4) $$\frac{\partial \Phi_n}{\partial \rho_{12}} = \int_{a_1}^{\infty} dx_1 \int_{a_2}^{\infty} dx_2 \cdots \int_{a_n}^{\infty} dx_n \frac{\partial^2 g_n}{\partial x_1 \partial x_2}$$

$$= \int_{a_3}^{\infty} dx_3 \int_{a_4}^{\infty} dx_4 \cdots \int_{a_n}^{\infty} dx_n g_n(a_1, a_2, x_3, \ldots, x_n; \Sigma_1) \geq 0,$$

and similarly,

(1.3.5) $$\frac{\partial \Phi_n}{\partial \rho_{ij}} \geq 0, \quad i \neq j.$$

If $\Sigma_2 = \{\kappa_{ij}\}$ is positive definite, set

(1.3.6) $$\gamma_{ij} = \lambda \rho_{ij} + (1-\lambda)\kappa_{ij}$$

then, $\{\gamma_{ij}\}$ is also positive definite for all λ satisfying $0 \leq \lambda \leq 1$.

(1.3.7) $$\frac{d\Phi_n}{d\lambda} = \sum_{i<j} \left(\frac{\partial \Phi_n}{\partial \gamma_{ij}}\right)\left(\frac{d\gamma_{ij}}{d\lambda}\right)$$

$$= \sum_{i<j} \left(\frac{\partial \Phi_n}{\partial \gamma_{ij}}\right)(\rho_{ij} - \kappa_{ij}),$$

and if $\rho_{ij} \geq \kappa_{ij}$, $\frac{\partial \Phi_n}{\partial \lambda} \geq 0$ since $\frac{\partial \Phi_n}{\partial \gamma_{ij}} \geq 0$. We get the required result.

Note that a slightly more general form of (1.3.3) has been given by Patil and Boswell (1970) as follows:

$$\frac{\partial g_n}{\partial \rho_{ij}} = \frac{\partial^2 g_n}{\partial x_i \partial x_j}\left(1 - \frac{\delta_{ij}}{2}\right), \quad i,j = 1,2,\ldots,n,$$

where $\delta_{ij} = 1$, if $i = j$ and $= 0$, if $i \neq j$.

In the special case of a correlation matrix $\Sigma_0 = \{\rho_{ij}\}$ with $\rho_{ij} = \alpha_i \alpha_j$ ($i \neq j$) and $-1 < \alpha_i < 1$, $i = 1,2,\ldots,n$, we can evaluate $\Phi_n(h_1,\ldots,h_n; \Sigma_0)$ as follows. The variates $X_1,\ldots,X_n$ can be represented via n+1 independent standard normal variates $Z_1, Z_2,\ldots,Z_n$; Y by the transformation

(1.3.8) $$X_i = (1-\alpha_i^2)^{\frac{1}{2}} Z_i + \alpha_i Y$$

and it follows that

(1.3.9) $$\Phi_n(h_1,\ldots,h_n; \Sigma_0) = \int_{-\infty}^{\infty} \left[\prod_{i=1}^{n} \Phi\left(\frac{h_i - \alpha_i y}{(1-\alpha_i^2)^{\frac{1}{2}}}\right)\right] d\Phi(y),$$

where Φ is the cdf of the standard normal random variables. The relation (1.3.9) was obtained by Dunnett and Sobel (1955) and in a more generalized form by Das (1956). For some special cases it is also given by Moran (1956), Ruben (1954), Gupta (1956),

Stuart (1958), and Ihm (1959). If $\rho_{ij} = \rho (\geq 0)$ for all i,j, then (1.3.9) reduces to

(1.3.10) $\quad \Phi_n(h,h,\ldots,h; \Sigma_0) = \int_{-\infty}^{\infty} \Phi^n[(h+\rho^{\frac{1}{2}}y)/(1-\rho)^{\frac{1}{2}}] d\Phi(y)$.

For values of h satisfying

(1.3.11) $\quad\quad\quad\quad\quad\quad\quad\quad \Phi_n(h,h,\ldots,h; \Sigma_0) = 1-\alpha$,

when α = 0.01, 0.025, 0.050, 0.100, 0.250; n = 1(1)10(2)50 and ρ = 0.100, 0.125, 0.200, $\frac{1}{3}$, 0.375, 0.400, $\frac{1}{2}$, 0.600, 0.625, $\frac{2}{3}$, 0.700, 0.750, 0.800, 0.875, 0.900; a complete table has been given by Gupta, Nagel and Panchapakesan (1973).

Note that the integral in (1.3.10) is the probability integral of the largest of n correlated normal variates N(0,1) with equal correlation ρ. Earlier Gupta (1963) gave a table of the function $\Phi_n(h,h,\ldots,h; \Sigma_0)$ for h = -3.50(0.10)3.50 and the same values of ρ as given above.

Dunnett and Sobel (1955) point out that if the matrix $\Sigma = \{\rho_{ij}\}$ has the structure $\rho_{ij} = \alpha_i \alpha_j$ $(i \neq j)$, where $-1 < \alpha_i < 1$ $(i = 1,2,\ldots,n)$, then Σ is guaranteed to be positive definite, since the associated quadratic form

$$\sum_{i=1}^{n} (1-\alpha_i^2)x_i^2 + (\sum_{i=1}^{n} \alpha_i x_i)^2$$

is positive for all $(x_1,\ldots,x_n) \neq (0,\ldots,0)$.

Next, we shall deal with the more complicated "two-sided" analogue of the above problem, namely, whether the probability

$$P\{|X_1| \leq c_1,\ldots,|X_n| \leq c_n\}$$

is a nondecreasing function of the correlations. It is shown that this is true in the important special case where the correlations are of the form $\lambda_i \lambda_j \rho_{ij}$, where $\{\rho_{ij}\}$ is some fixed correlation matrix. A precise result of this nature was obtained by Šidák (1968): If $R(\lambda)$ denotes the correlation matrix with $\rho_{1j}(\lambda) = \rho_{j1}(\lambda) = \lambda \rho_{1j}$ for $j > 1$, $0 \leq \lambda \leq 1$; $\rho_{ij}(\lambda) = \rho_{ij}$ for all other i,j and P_λ denotes the probability distribution corresponding to $R(\lambda)$ then

(1.3.12) $\quad\quad\quad\quad\quad\quad \frac{d}{d\lambda} P_\lambda [|X_i| \leq c_i, 1 \leq i \leq n] \geq 0$.

Šidák's proof of (1.3.12) is very lengthy and from his remarks it seems as if Slepian's method is not applicable for this "two-sided" version. Jogdeo (1970), on the other hand, has shown that Slepian's method combined with a lemma given in the following (see Lemma 1.3.4), readily lead to the desired inequality. This lemma is a simple corollary to the fundamental inequality of Anderson (1955) which also served as the key to Šidák's proof. As observed by Šidák it suffices to prove (1.3.12) under the assumption that $R = \{\rho_{ij}\}$ is positive definite.

Note that the sufficient conditions of the lemma which was proved by Jogdeo (1970) (see Lemma 1.3.4) are not enough to prove it. We shall modify the conditions and give a proof in detail. For this purpose, we introduce Anderson's Fundamental Lemma (1950) as follows.

If one has a function f(x) on the real line which is symmetric about 0 and unimodal (i.e., $f(kx) \geq f(x)$, $0 \leq k \leq 1$), it is obvious that the integral of f(x) over an interval of fixed length is maximized if the interval is centered at origin; in fact, the integral is a nonincreasing function of the distance of the midpoint from the origin. A direct result of this is that if a random variable X has density f(x) which is symmetric about the origin and unimodal and Y is an another random variable independently of X, then

$$P\{|X| \leq a\} \geq P\{|X+Y| \leq a\}.$$

Anderson (1955) has generalized the results to n-dimensional Euclidean space. The interval is replaced by a symmetric (wrt $\underline{0}$) convex set; the condition of unimodality is expressed by the condition that the set of points for which the function is at least equal to a given value is convex.

As an application to interval estimation problem, consider a random sample $X_1,\ldots,X_n$ with common normal distribution $N(\mu,1)$. The usual confidence interval for μ is $|\bar{X}-\mu| \leq c$, which is a symmetric region about μ. Since the distribution of $\bar{X}-\mu$ is symmetric about the origin and unimodal, it follows that the above confidence interval has the maximum confidence level among all intervals of fixed length 2c.

We shall now state and prove Anderson's result (1955).

1.3.2 <u>Lemma</u>. Let E be a convex set in n-dimensional Euclidean space, symmetric about the origin. Let $f_n(\underline{x}) \geq 0$ be a function such that (i) $f_n(\underline{x}) = f_n(-\underline{x})$ (ii) $\{\underline{x} | f_n(\underline{x}) \geq u\} = K_u$ is convex for every u, $(0 < u < \infty)$, and (iii) $\int_E f_n(\underline{x})d\underline{x} < \infty$ (in the Lebesgue sense). Then

(1.3.13) $$\int_E f_n(\underline{x}+k\underline{y})d\underline{x} \geq \int_E f_n(\underline{x}+\underline{y})d\underline{x}$$

for $0 \leq k \leq 1$.

<u>Proof</u>. An equivalent way of writing (1.3.13) is

$$\int_{E+k\underline{y}} f_n(\underline{x})d\underline{x} \geq \int_{E+\underline{y}} f_n(\underline{x})d\underline{x},$$

where E+y is the set E translated by the vector $\underline{y}$. The result follows almost directly after we prove that, for every u,

$$V\{(E+k\underline{y}) \cap K_u\} \geq V\{(E+\underline{y}) \cap K_u\},$$

where V{ } indicates the volume of the set.

Let $\alpha[(E+\underline{y}) \cap K_u] + (1-\alpha)[(E-\underline{y}) \cap K_u]$ denote the set obtained by taking all linear combinations $\alpha\underline{z} + (1-\alpha)\underline{w}$, where $\underline{z} \in (E+\underline{y}) \cap K_u$ and $\underline{w} \in (E-\underline{y}) \in K_u$ and $0 \leq \alpha \leq 1$. Let $\alpha = \frac{1+k}{2}$, so that $\alpha\underline{y} + (1-\alpha)(-\underline{y}) = k\underline{y}$. Then

$$(E+k\underline{y}) \cap K_u \supset \alpha[(E+\underline{y}) \cap K_u] + (1-\alpha)[(E-\underline{y}) \cap K_u],$$

because K_u is convex and

$$E+k\underline{y} \supset \alpha(E+\underline{y}) + (1-\alpha)(E-\underline{y})$$
$$= [\alpha E+(1-\alpha)E] + k\underline{y}.$$

Thus
$$V\{(E+k\underline{y}) \cap K_u\} \geq V\{\alpha[(E+\underline{y}) \cap K_u] + (1-\alpha)[(E-\underline{y}) \cap K_u]\}.$$

$(E+\underline{y}) \cap K_u$ is the mirror image through the origin of $(E-\underline{y}) \cap K_u$, and therefore these two sets have the same volume. Then
$$V\{\alpha[(E+\underline{y}) \cap K_u] + (1-\alpha)[(E-\underline{y}) \cap K_u]\}$$
$$\geq V\{(E+\underline{y}) \cap K_u\}$$

by the Brunn-Minkowski Theorem (Bonnesen and Fenchel (1948)), which states that
$$V^{\frac{1}{n}}\{(1-\theta)E_0+\theta E_1\} \geq (1-\theta)V^{\frac{1}{n}}(E_0)+\theta V^{\frac{1}{n}}(E_1),$$

(E_0 and E_1 nonempty, $0 \leq \theta \leq 1$). Thus
$$V\{(E+k\underline{y}) \cap K_u\} = H(u) \geq V\{(E+\underline{y}) \cap K_u\} = H^*(u).$$

Properties of the Lebesgue and Lebesgue-Stieltjes integrals lead to
$$\int_{E+k\underline{y}} f_n(\underline{x})d\underline{x} - \int_{E+\underline{y}} f_n(\underline{x})d\underline{x}$$
$$= -\int_0^\infty u dH(u) + \int_0^\infty u dH^*(u) = \int_0^\infty u d[H^*(u)-H(u)].$$

Integration by parts shows

(1.3.14)
$$\int_a^b u d[H^*(u)-H(u)] = b[H^*(b)-H(b)]$$
$$- a[H^*(a)-H^*(a)] + \int_a^b [H(u)-H^*(u)]du.$$

Since $f_n(\underline{x})$ has a finite integral over E, $bH(b) \to 0$ as $b \to \infty$ and hence also $bH^*(b) \to 0$ as $b \to \infty$; therefore the first term on the right in (1.3.14) can be made arbitrarily small in absolute value. If $a \geq 0$, the second term above is nonnegative as well as the third. Thus

$$\int_0^\infty u d[H^*(u) - H(u)] \geq 0.$$

The proof is complete.

Note that the integral
$$\phi(\underline{y}) = \int_E f_n(\underline{x}+\underline{y})d\underline{x}$$

is a symmetric function and is unimodal in the sense that along a given ray through the origin the integral is a nondecreasing function of the distance from the origin. However, $\phi(\underline{y})$ does not necessarily have the unimodality of $f_n(\underline{x})$; that is, $\{\underline{y}|\phi(\underline{y}) \geq u\}$ is not necessarily convex. As an example (see Sherman (1955)), take the case $n = 2$, and

$$f(\underline{x}) = \begin{cases} 3, & |x_1| \le 1, |x_2| \le 1, \\ 2, & |x_1| \le 1, 1 < |x_2| \le 5, \\ 0, & \text{otherwise}, \end{cases}$$

where $\underline{x} = (x_1, x_2)$. Let E be the set of vectors where $|x_1| \le 1$, $|x_2| \le 1$. The set $\{\underline{x} | \phi(\underline{x}) \ge 6\}$ is not convex since for $\underline{x} = (0.5, 4)$ and $\underline{x} = (1, 0)$, $\phi(\underline{x}) = 6$, while for $\underline{x} = (0.75, 2)$, $\phi(\underline{x}) < 6$.

The following important result of Anderson (1955) is an immediate consequence of his fundamental lemma.

1.3.3 **Corollary.** Let $\underline{X}$ be normally distributed with mean $\underline{0}$ and covariance matrix Σ; Let $\underline{Z}$ be normally distributed with mean $\underline{0}$ and covariance matrix ψ, where $\psi - \Sigma$ is positive semi-definite. If E is a convex set, symmetric about the origin, then $P\{\underline{X} \in E\} \ge P\{\underline{Z} \in E\}$. If $h(\underline{x})$ is a symmetric function such that $\{\underline{x} | h(\underline{x}) \le v\}$ is convex, then $P\{h(\underline{X}) \le v\} \ge P\{h(\underline{Z}) \le v\}$. If E is a bounded set and $\psi - \Sigma \ne \underline{0}$, then strict inequality holds.

Proof. Let $\underline{Y}$ be normally distributed with mean $\underline{0}$ and covariance matrix $\psi - \Sigma$, and let $\underline{Y}$ be independent distributed with $\underline{X}$. Then $\underline{Z}$ has the same distribution as $\underline{X} + \underline{Y}$, and it follows from Lemma 1.3.2.

Returning to the two-sided problem, let g_n be the multivariate normal density of $X_1, \ldots, X_n$ and $\underline{\rho}_{12}$ be the correlation (column) vector between X_1 and $\underline{X}_2 = (X_2, \ldots, X_n)$, (for n = 2, $\underline{\rho}_{12} = \rho_{12}$ is the correlation coefficient between X_1 and X_2). To motivate the basic idea of the proof, we use (1.3.3) to obtain

$$(1.3.15) \quad \frac{d}{d\rho_{12}} P[|X_1| \le c_1, |X_2| \le c_2]$$

$$= 2\{g_2(c_1, c_2; \rho_{12}) - g_2(-c_1, -c_2; \rho_{12})\}$$

$$\ge 0 \ (\le 0) \quad \text{for } \rho_{12} > 0 \ (< 0),$$

which has an interpretation: the derivative in (1.3.15) is nonnegative in the direction of ρ_{12}, i.e.

$$\rho_{12} \frac{d}{d\rho_{12}} P[|X_1| \le c_1, |X_2| \le c_2] \ge 0.$$

From this point of view, the inequality (1.3.12) is the same as the proposition: the derivative of $P[|X_i| \le c_i, i = 1, 2, \ldots, n]$ with respect to $\underline{\rho}_{12}$ is nonnegative in the direction of $\underline{\rho}_{12}$. Jogdeo (1970) has shown that this is a consequence of the following lemma. Note that we have modified the sufficient conditions and prove it as follows.

1.3.4 **Lemma.** Let E be a convex, symmetric set in n-dimensional Euclidean space and let f_n be a nonnegative function satisfying

(i) $f_n(\underline{x}) = f_n(-\underline{x})$,

(ii) f_n is differentiable and for each j, $1 \leq j \leq n$,
$$\lim_{k \to 1} \int_E |D_j f_n(\underline{x}+\underline{b}+(k-1)\underline{c})| d\underline{x} = \int_E |D_j f_n(\underline{x}+\underline{b})| d\underline{x} < \infty,$$

(iii) f_n is unimodal: $K_u = \{\underline{x} | f_n(\underline{x}) \geq u\}$ is convex for every $u > 0$,

(iv) $\int_E f_n(\underline{x}) d\underline{x} < \infty$.

Then, for any arbitrary but fixed (column) vector $\underline{a}$,
$$\int_E \underline{a}' \frac{d}{d\underline{x}} f_n(\underline{x}+\underline{a}) d\underline{x} \leq 0,$$

where $\underline{a}'$ denotes the transpose of $\underline{a}$, $\underline{c}$ and $\underline{b}$ are any constant vectors and D_j is the partial derivative with respect to the jth component.

<u>Proof</u>. Under the conditions of Lemma 1.3.2, $\int_E f_n(\underline{x}+k\underline{a}) d\underline{x}$ is a nonincreasing function of k, where $0 \leq k \leq 1$. For $0 < k < 1$,

$$0 \leq \int_E [f_n(\underline{x}+k\underline{a}) - f_n(\underline{x}+\underline{a})] d\underline{x}$$

$$= (k-1) \int_E \frac{1}{k-1} \sum_{j=1}^{n} [f_n(\underline{x}+\underline{a}+\underline{v}_j) - f_n(\underline{x}+\underline{a}+\underline{v}_{j-1})] d\underline{x},$$

where $\underline{v}_0 = \underline{0}$, $\underline{v}_j = (k-1)a_1\underline{e}_1 + \ldots + (k-1)a_j\underline{e}_j$, $\underline{e}_j$ is the vector in n-dimensional Euclidean space whose ith coordinate is 1 and whose other coordinates are all 0, and for each j, $(1 \leq j \leq n)$, $a_j > 0$, say,

$$\lim_{k \to 1} \sup \int_E \left| \frac{1}{k-1} [f_n(\underline{x}+\underline{a}+\underline{v}_j) - f_n(\underline{x}+\underline{a}+\underline{v}_{j-1})] \right| d\underline{x}$$

$$= \lim_{k \to 1} \sup \int_E \frac{1}{1-k} \left| \int_{(k-1)a_j}^{0} D_j f_n(\underline{x}+\underline{a}+\underline{v}_{j-1}+y\underline{e}_j) dy \right| d\underline{x}$$

$$\leq \lim_{k \to 1} \sup \frac{1}{1-k} \int_{(k-1)a_j}^{0} \int_E |D_j f_n(\underline{x}+\underline{a}+\underline{v}_{j-1}+y\underline{e}_j)| d\underline{x} dy$$

$$= \int_E a_j |D_j f_n(\underline{x}+\underline{a})| d\underline{x} < \infty,$$

and a similar result may be obtained for $a_j < 0$, hence

$$0 \geq \lim_{k \to 1} \int_E \frac{1}{k-1} \sum_{j=1}^{n} [f_n(\underline{x}+\underline{a}+\underline{v}_j) - f_n(\underline{x}+\underline{a}+\underline{v}_{j-1})] d\underline{x}$$

$$= \lim_{k \to 1} \int_E \sum_{j=1}^{n} a_j D_j f_n(\underline{x}+\underline{a}+\underline{v}_{j-1}+(k-1)a_j\theta\underline{v}_j) d\underline{x}, \text{ for some } \theta, 0 < \theta < 1,$$

$$= \int_E \sum_{j=1}^{n} a_j D_j f_n(\underline{x}+\underline{a}) d\underline{x},$$

i.e.
$$\int_E \underline{a}' \frac{d}{d\underline{x}} f_n(\underline{x}+\underline{a})d\underline{x} \leq 0.$$

From Lemma 1.3.4, Jogdeo (1970) has simplified the proof of Sidák's theorem (1968) as follows.

1.3.5 Theorem. Let $(X_1,\ldots,X_n)$ be a random vector having an n-variate normal distribution with mean values 0, variances 1, and with the positive definite correlation matrix $R(\lambda) = \{\rho_{ij}(\lambda)\}$ depending on a parameter λ, $0 \leq \lambda \leq 1$, in the following way: under the probability law P_λ, we have $\rho_{1j}(\lambda) = \rho_{j1}(\lambda) = \lambda\rho_{1j}$ for $j \geq 2$, $\rho_{ij}(\lambda) = \rho_{ji}(\lambda) = \rho_{ij}$ for $i,j \geq 2$, $i \neq j$, where $\{\rho_{ij}\}$ is some fixed correlation matrix. Then

(1.3.16) $$P(\lambda) = P_\lambda\{|X_1| \leq c_1, \ldots, |X_n| \leq c_n\}$$

is a nondecreasing function of λ, $0 \leq \lambda \leq 1$, for any positive fixed numbers $c_1, c_2, \ldots, c_n$.

Proof. As remarked before, the inequality (1.3.12) is equivalent to the directional derivative being nonnegative:

$$\rho'_{12} \frac{d}{d\rho_{12}} P[|X_i| \leq c_i, \ i = 1,2,\ldots,n] \geq 0.$$

Since,

$$P[|X_i| \leq c_i, \ i = 1,2,\ldots,n] = 2\int_0^{c_1} \int_{\underline{c}_2} g_n(x_1,\underline{x}_2; R) dx_1 d\underline{x}_2,$$

where $R = \{\rho_{ij}\}$ and $\underline{c}_2 = [-c_2,c_2]\times[-c_3,c_3]\times\ldots\times[-c_n,c_n]$, using (1.3.3) it follows that

(1.3.17) $$\rho'_{12} \frac{d}{d\rho_{12}} P[|X_i| \leq c_i, \ i = 1,2,\ldots,n]$$

$$= 2\rho'_{12} \int_0^{c_1} \int_{\underline{c}_2} \frac{\partial^2}{\partial x_1 \partial \underline{x}_2} g_n(x_1,\underline{x}_2; R) dx_1 d\underline{x}_2$$

$$= 2\rho'_{12}[\int_{\underline{c}_2} \frac{\partial}{\partial \underline{x}_2} g_n(c_1,\underline{x}_2; R) d\underline{x}_2 - \int_{\underline{c}_2} \frac{\partial}{\partial \underline{x}_2} g_n(0,\underline{x}_2; R) d\underline{x}_2].$$

It is easy to verify that the second term inside the rectangular brackets on the right hand side of the last equality sign in (1.3.17) is zero. Further, the first term can be rewritten by using the conditional density function of $\underline{x}_2$ given $X_1 = c_1$. Thus

(1.3.18) $$\rho'_{12} \frac{d}{d\rho_{12}} P[|X_i| \leq c_i, \ i = 1,2,\ldots,n]$$

$$= b\rho'_{12} \int_{\underline{c}_2} \frac{\partial}{\partial \underline{x}_2} \psi(\underline{x}_2-c_1\rho_{12}; R_1) d\underline{x}_2,$$

where b is a positive constant, ψ is the conditional density of $\underline{x}_2$ given $\underline{x}_1$ and R_1 is the conditional variance covariance matrix. Since R_1 is positive definite, ψ satisfies the conditions of Lemma 1.3.4 and thus the right hand side of (1.3.18) when multiplied by $-c_1$ is seen to be nonpositive. Hence the left hand side of (1.3.18) must be nonnegative. This completes the proof.

Of course, in Theorem 1.3.5, we may permute the subscripts, and write any other subscript in place of 1. In this manner, we can restate Theorem 1.3.5 as follows.

1.3.6 Corollary. Let $(X_1, X_2, \ldots, X_n)$ be as stated in Theorem 1.3.5, except that its correlation matrix, depending on n parameters $\lambda_1, \lambda_2, \ldots, \lambda_n$ ($0 \leq \lambda_i \leq 1$, $1 \leq i \leq n$) be given, under the probability law $P_{\lambda_1, \lambda_2, \ldots, \lambda_n}$, by $\{\lambda_i \lambda_j \rho_{ij}\}$ for $i \neq j$. Then

$$P(\lambda_1, \lambda_2, \ldots, \lambda_n) = P_{\lambda_1, \ldots, \lambda_n}\{|X_1| \leq c_1, \ldots, |X_n| \leq c_n\}$$

is a nondecreasing function of each λ_i, $0 \leq \lambda_i \leq 1$, $i = 1, 2, \ldots, n$.

In applications, we often encounter the multivariate analogue of Student's t-distribution. Recall that this is the distribution of $(X_1/s, \ldots, X_n/s)$ where $(X_1, \ldots, X_n)$ have a multivariate normal distribution with zero means, common variance σ^2, and some correlation matrix R, and where $\frac{\nu s^2}{\sigma^2}$ has the χ^2-distribution with ν-degrees of freedom, which is distributed independently of $(X_1, \ldots, X_n)$ (s^2 is the usual pooled sample variance). It is easy to see that the results of Slepian and Šidák continue to hold also for the multivariate analogue of Student t-distribution. The proof is immediate if we use the conditional distribution given s, and then finally take the expectation.

Intuitively, in view of Slepian's theorem (1962), the values of correlations may be regarded as some measures of how much the variables "hang together". Roughly speaking, the larger the (positive) correlations are, the more the individual variables "hang together", and the more likely it is that they will behave similarly. From this point of view, Slepian's inequality (1.3.2) can be regarded as a confirmation of this intuitive principle, and one might also then perhaps expect that an analogue of this inequality will hold for the case of a two-sided barrier (at least for positive correlations). However, Šidák (1968) gives a counterexample to show that this is not true and such a general analogue does not hold as follows.

The example concerns the three dimensional case, n = 3, and (X_1, X_2, X_3) have a normal distribution with mean values 0, variances 1, and some correlation matrix $\{\sigma_{ij}\}$.

1.3.7 Example. Šidák (1968) asserts that if $\sigma_{13} - \sigma_{12}\sigma_{23} < 0$ (> 0), and if c_2 is sufficiently small, then

(1.3.19)
$$\frac{\partial}{\partial \sigma_{13}} P\{|X_1| \leq c_1, |X_2| \leq c_2, |X_3| \leq c_3\}$$

$$< 0 \ (> 0, \text{respectively}).$$

Thus, in the first case, if σ_{13} increases, the probability in question decreases.

Furthermore, Das Gupta, Eaton, Olkin, Perlman, Savage, and Sobel (1970) extend the results from normal to elliptically contoured density of the form

$$|\Sigma|^{-\frac{1}{2}} f(\underline{x}'\Sigma^{-1}\underline{x}),$$

where $\int_0^\infty r^{p-1} f(r) dr < \infty$, $\underline{x}' = (x_1, \ldots, x_p)$ and $\Sigma = \{\sigma_{ij}\}$ is a positive definite matrix. The normal distribution is clearly a special case.

1.4. Dependence and Association of Random Variables

We shall discuss here some concepts for the monotonicity property of distributions which are more general than the results discussed in Section 1.3. Usually, the problems involving dependent pairs of variables (X,Y) have been studied most intensively in the cases of bivariate normal distributions and 2x2 tables. This is due primarily to the importance of these cases but perhaps partly also to the fact that they exhibit a particularly simple form of dependence. Studies involving the general case center mainly around two problems:

(i) tests of independence;
(ii) definition and estimation of measures of association.

In most treatments of these problems, there occurs implicitly a concept which is of importance also in the evaluation of the performance of certain multiple decision procedures, namely, the concept of positive (or negative) dependence or association. The concepts of dependence and association are also very important in reliability theory.

For a first definition, we compare the probability of any quadrant $X \leq x$, $Y \leq y$ under the distribution F of (X,Y) with the corresponding probability in the case of independence. Lehmann (1966) defines the following definitions.

1.4.1 <u>Definition</u>. The pair (X,Y) or its distribution F is positively (negatively) quadrant dependent if

(1.4.1) $\qquad P(X \leq x, Y \leq y) \geq (\leq) P(X \leq x) P(Y \leq y)$

for all x,y. The dependence is strict if inequality holds for at least one pair (x,y).

In the sequel, let us denote the family of all distributions F satisfying (1.4.1) by $\mathcal{F}_1(\mathcal{G}_1)$. To simplify the notation we shall write $(X,Y) \in \mathcal{F}_1$ to indicate that the distribution of (X,Y) belongs to $\mathcal{F}_1$.

Lehmann (1966) proves an important result for the relation between the distributions and their expectations as follows.

1.4.2 <u>Theorem.</u> If F denotes the joint and F_X and F_Y the marginal distributions of X and Y, then

(1.4.2) $E(XY) - EXEY = \int_{-\infty}^{\infty} \int_{-\infty}^{\infty} [F(x,y) - F_X(x)F_Y(y)] dx dy$

provided the expectations on the left hand side exist.

Proof. Let (X_1, Y_1), (X_2, Y_2) be independent, each distributed according to F. Then

$$2[E(X_1 Y_1) - EX_1 EY_1] = E[(X_1 - X_2)(Y_1 - Y_2)]$$

$$= E\int_{-\infty}^{\infty} \int_{-\infty}^{\infty} [I(u, X_1) - I(u, X_2)][I(v, Y_1) - I(v, Y_2)] du dv$$

where $I(u,x) = 1$ if $u \leq x$ and $= 0$ otherwise. Since $E|XY|$, $E|X|$ and $E|Y|$ are assumed finite, we can take the expectation under integral sign, and the resulting expression is seen to reduce to twice the right hand side of (1.4.2). This completes the proof.

Note that it is easy to see if $(X,Y) \in \mathcal{F}_1$ then $cov(X,Y) \geq 0$ by using Theorem 1.4.2. For any pair of random variables (X,Y) with bivariate normal distribution which has positive correlation coefficient is in $\mathcal{F}_1$. Lehmann (1966) also shows that if $(X,Y) \in \mathcal{F}_1$ then Kendall's τ, Spearman's ρ_s and the quadrant measure q discussed by Blomquist (1950) are all non-negative.

For the purpose of relating the positive quadrant dependence to other useful concepts, we rewrite (1.4.1) as

(1.4.3) $P(Y \leq y | X \leq x) \geq P(Y \leq y)$

and in this form clearly expresses the fact that the knowledge of X being small increases the probability of Y being small. This is a very important idea which is often used in reliability theory. It may be felt that the intuitive concept of positive dependence is better represented by the stronger condition

(1.4.4) $P(Y \leq y | X \leq x) \geq P(Y \leq y | X \leq x')$

for all $x < x'$ and all y. We recall that Y is strongly positive quadrant dependent on X. Rather than (1.4.4), we shall consider the still stronger condition

(1.4.5) $P(Y \leq y | X = x)$ is nonincreasing in x,

which was discussed by Tukey (1958) and Lehmann (1959). We state (1.4.5) as the following definition.

1.4.3 Definition. If $P(Y \leq y | X = x)$ is nonincreasing (or nondecreasing) in x, we shall say that Y is positively (or negatively) regression dependent on X; the family of all distributions F of (X,Y) with the property (1.4.5) will be denoted by $\mathcal{F}_2$ (or $\mathcal{G}_2$).

Lehmann (1966) gives an example to show $\mathcal{F}_2$ (or $\mathcal{G}_2$) exists.

1.4.4 Example. Let $Y = \alpha + \beta X + U$, where X and U are independent. Then Y is positively or negatively regression dependent on X as $\beta \geq 0$ or ≤ 0. This is obvious since the conditional distribution of Y given X=x is that of $\alpha + \beta x + U$ and hence is clearly stochastically increasing in x if $\beta > 0$. In particular, it follows from this example that the components of a bivariate normal distribution are positively or negatively regression dependent as $\rho \geq 0$ or $\rho \leq 0$.

The relation between $\mathcal{F}_1$ and $\mathcal{F}_2$ has been shown below by Lehmann (1966).

1.4.5 <u>Theorem</u>. The condition (1.4.3) to (1.4.5) are connected by the implications (1.4.5) $\Rightarrow$ (1.4.4) $\Rightarrow$ (1.4.3) and hence $\mathcal{F}_2 \subset \mathcal{F}_1$.

<u>Proof</u>. It is obvious to show (1.4.4) $\Rightarrow$ (1.4.3) by taking $x' = \infty$. To show that (1.4.5) $\Rightarrow$ (1.4.4) let

$$h_y(u) = P(Y \leq y | X = u),$$

so that for any y,

$$P(Y \leq y, X \leq x) = \int_{-\infty}^{x} h_y(u) dF_X(u),$$

where F_X denotes the marginal distribution of X. Under the assumption that h_y is non-increasing, we must therefore show that

$$\int_{-\infty}^{x} h_y(u) dF_X(u)/P(X \leq x) \geq \int_{-\infty}^{x'} h_y(u) dF_X(u)/P(X = x'), \text{ for } x < x',$$

which can be proved by differentiation.

While in many standard examples of positive dependence, the distributions satisfy not only (1.4.3) but also (1.4.5), the second of these conditions is much stronger than the first. The following are some examples, in which there is a strong intuitive feeling of positive or negative dependence, where there is quadrant dependence, but where the stronger condition of regression dependence is not satisfied.

1.4.6 <u>Example</u>. (Lehmann (1966)). Consider testing the nonparametric hypothesis of the equality of s distributions $H: F_1 = \ldots = F_s$ against the alternative K_i that F_i has slipped to the right, on the basis of samples X_{ik} ($k = 1, \ldots, n_i$) from F_i. If the $N = \sum_{i=1}^{s} n_i$ observations are ranked, and R_{ik} denotes the rank of X_{ik} and $R_i = \sum_{k=1}^{n_i} R_{ik}$, a distribution free test rejects H in favor of K_i when R_i is sufficiently large. In this connection Doornbos and Prins (1958) give a proof by the fact that $(R_i, R_j) \in \mathcal{G}_1$. An examination of the proof shows that it carries over to the case in which an arbitrary set of numbers $v_1, \ldots, v_N$ (rather than the set of integers $1, 2, \ldots, N$) is divided at random into s groups $n_1, \ldots, n_s$ elements respectively, with R_i denoting the sum of the v-values in the ith group. As was stated above, the variables (R_i, R_j) are in $\mathcal{G}_1$ for all values $v_1, \ldots, v_N$. While they are also in $\mathcal{G}_2$ for the special case that the v's are all zero or one, this is not true in general, as is seen by putting $v_1 = 3$, $v_2 = 3$, $v_3 = 5$, $v_4 = 15$; $n_1 = 2$, $n_2 = 1$; and $s = 2$. It is easily checked that in this case $P(R_2 \leq 1 | R_1 \leq x)$ is not increasing in x, and that therefore $X = R_1$, $Y = R_2$ do not even satisfy (1.4.4).

This example also demonstrates the asymmetry in x, y of (1.4.4) and (1.4.5).

For while $P\{R_2 \leq y | R_1 \leq x\}$ and $P\{R_2 \leq y | R_1 = x\}$ are not increasing functions of x for all y, $P\{R_1 \leq x | R_2 \leq y\}$ and $P\{R_1 \leq x | R_2 = y\}$ are nondecreasing functions of y for all x. In fact, consider any $v_1,\ldots,v_N$ from which we draw first a sample of size n_1 with sample sum R_1 and then a sample of size $n_2 = 1$ with sample value R_2. Then $P\{R_1 \leq x | R_2 = y\}$ is an increasing function of y. This is easily seen by constructing from a random variable with the distribution of R_1 given y, a variable with the distribution of R_1 given $y' > y$, which is always at least as large as the first variable and with positive probability is actually larger.

1.4.7 <u>Example</u>. (Lehmann (1966)). If U, V are independent then X = U+V, Y=U belong to $\mathfrak{F}_1$. That they do not necessarily satisfy (1.4.4) or (1.4.5) is shown by the case in which both variables take on the values 0, 2, 3 with probabilities p, q, r, (p + q + r = 1). For it is then easily checked that
$$P(Y \leq 2 | X \leq 3) < P(Y \leq 2 | X \leq 4).$$

Lehmann (1966) also discussed an example to show that (X,Y) satisfies (1.4.4) but not (1.4.5) as follows.

1.4.8. <u>Example</u>. Let the distribution of (X,Y) be given by the following 3x3 table

		y		
		1	2	3
	1	p	0	0
x	2	0	q	0
	3	s	0	r

Then it is easily checked that (X,Y) satisfies (1.4.4) provided
(1.4.6) $\qquad\qquad\qquad qs \leq pr;$
On the other hand, (X,Y) satisfies (1.4.5) only if s = 0.

Condition (1.4.4) which defined regression dependence requires the conditional variable Y given x to be stochastically increasing. An even stronger condition is obtained by requiring the conditional density of Y given x to have monotone likelihood ratio. Lehmann (1966) then says that (X,Y) or its distribution F shows positive likelihood ratio dependence.

1.4.9 <u>Definition</u>. If f(x,y) is the joint density of X and Y with respect to some product measure, and satisfies
(1.4.7) $\qquad\qquad f(x,y')f(x',y) \leq f(x,y)f(x',y')$
for all $x < x'$, $y < y'$, then we say that (X,Y) or its distribution F shows positive likelihood ratio dependence (i.e., MLR). The family of all distributions F satisfying this condition will be denoted by $\mathfrak{F}_3$. If the inequality of (1.4.7) is reversed, F is negatively likelihood ratio dependent and belongs to $\mathcal{G}_3$. Condition (1.4.7) is symmetric in X and Y.

The following are some examples with monotone likelihood ratio dependence

(Lehmann (1966)).

1.4.10. Example.
(i) A bivariate normal density is in $\mathcal{F}_3$ or $\mathcal{G}_3$ as the correlation coefficient $\rho \geq 0$ or ≤ 0.
(ii) Any two components of a multinomial distribution are in $\mathcal{G}_3$.

1.4.11 Example.
Consider the joint distribution F of any two dependent binary variables I_0 and I_1 with, say, $p_{ij} = P(I_0=i, I_1=j)$, $i = 0,1$; $j = 0,1$. (Such a distribution defines the random structure of a two by two table if I_0 and I_1 indicate the occurrence of the two characteristics in question.) Then it is seen that F is in $\mathcal{F}_3$ (and hence in $\mathcal{F}_1$ and $\mathcal{F}_2$) if

$$(1.4.8) \qquad p_{00}p_{11} \geq p_{01}p_{10}$$

and in $\mathcal{G}_3$ if the inequality is reversed. As in the normal case, all distributions in this family therefore show likelihood ratio dependence.

1.4.12 Example.
Let U, V be independently distributed with densities g and h respectively, and let X=U, Y = U+V. Then we saw earlier in Example 1.4.6 that Y is always positively regression dependent on X but not necessarily X on Y. Now the joint density of X and Y is $g(x)h(y-x)$ and conditions (1.4.7) therefore reduces to

$$(1.4.9) \qquad h(y-x')/h(y-x) \leq h(y'-x')/h(y'-x).$$

This condition is satisfied provided -log h is convex (see Lehmann (1959), p. 330). For such densities, (X,Y) belongs to $\mathcal{F}_3$ and hence not only (X,Y) but also (Y,X) belongs to $\mathcal{F}_2$. This last result was proved by Efron (1965).

1.4.13 Example.
Let $Z_1,\ldots,Z_n$ be independently distributed according to a univariate distribution G, and let X and Y be two order statistics of the Z's, say $X = Z^{(r)}$ and $Y = Z^{(s)}$ with $r < s$. Then the joint distribution F of X and Y has density

$$f(x,y) = [G(x)]^{r-1}[G(y)-G(x)]^{s-r-1}[1-G(y)]^{n-s},$$

for $x < y$, with respect to the product measure $G \times G$. This density satisfies (1.4.7) so that $F \in \mathcal{F}_3$. That $F \in \mathcal{F}_2$ and hence $\text{Cov}(Z^{(r)}, Z^{(s)}) \geq 0$ was proved by Bickel (1965).

1.4.14 Definition.
We say that the random variables $T_1,\ldots,T_n$ are associated if

$$(1.4.10) \qquad \text{Cov}[f(\underline{T}),g(\underline{T})] \geq 0,$$

where $\underline{T} = (T_1,\ldots,T_n)$, for all nondecreasing functions f and g for which $Ef(\underline{T})$, $Eg(\underline{T})$, $Ef(\underline{T})g(\underline{T})$ exist.

In particular, we know that for binary random variables X,Y association is equivalent to $\text{Cov}[X,Y] \geq 0$ by considering all possible binary nondecreasing functions $r(X,Y)$:

$$(r\equiv 0) \le (r=XY) \le \begin{Bmatrix} r=X \\ r=Y \end{Bmatrix} \le (r=X+Y-XY) \le (r\equiv 1).$$

After Esary, Proschan and Walkup (1967) state the previous fact, they continue to study some interesting results as follows:
(i) If T is positively regression dependent on S, then S and T are associated.
(ii) S and T are positively quadrant dependent if and only if $Cov[f(X),g(T)] \ge 0$ for all nondecreasing functions f,g.
(iii) For general random variables no two of the following conditions are equivalent:
(1.4.11) $Cov[S,T] \ge 0$,
(1.4.12) $Cov[f(S),g(T)] \ge 0$ for all nondecreasing functions f,g,
(1.4.13)   S,T associated.
It can be shown that
$$(1.4.13) \Rightarrow (1.4.12) \Rightarrow (1.4.11).$$
We can find some S,T satisfying (1.4.11) but not (1.4.12). To show S,T may satisfy (1.4.12) but not (1.4.13), let S,T take on values a_1, a_2, a_3 ($a_1 < a_2 < a_3$) with probabilities

	$S = a_1$	$S = a_2$	$S = a_3$
$T = a_3$	$\frac{8}{64}$	0	$\frac{15}{64}$
$T = a_2$	0	$\frac{18}{64}$	0
$T = a_1$	$\frac{15}{64}$	0	$\frac{8}{64}$

Finally, to show S,T may satisfy (1.4.13) but not (1.4.5), let S,T take on values $a_1 < a_2 < a_3$ with probabilities:

	$S = a_1$	$S = a_2$	$S = a_3$
$T = a_3$	$\frac{1}{8}$	0	$\frac{1}{4}$
$T = a_2$	0	$\frac{1}{4}$	0
$T = a_1$	$\frac{1}{4}$	0	$\frac{1}{8}$

However, for binary random variables X,Y, conditions (1.4.11), (1.4.12), (1.4.13) and (1.4.5) are equivalent. This is a consequence of the following two facts:
(1) As was pointed out in Example 1.4.11, binary X,Y satisfy (1.4.5) if and only if
(1.4.14) $P[X=0, Y=0]P[X=1, Y=1]$
            $\ge P[X=0, Y=1]P[X=1, Y=0].$
(2) Binary X,Y satisfy (1.4.11) if and only if (1.4.14) holds, as may be verified directly.
Finally, let us note that we have proved the following diagram:

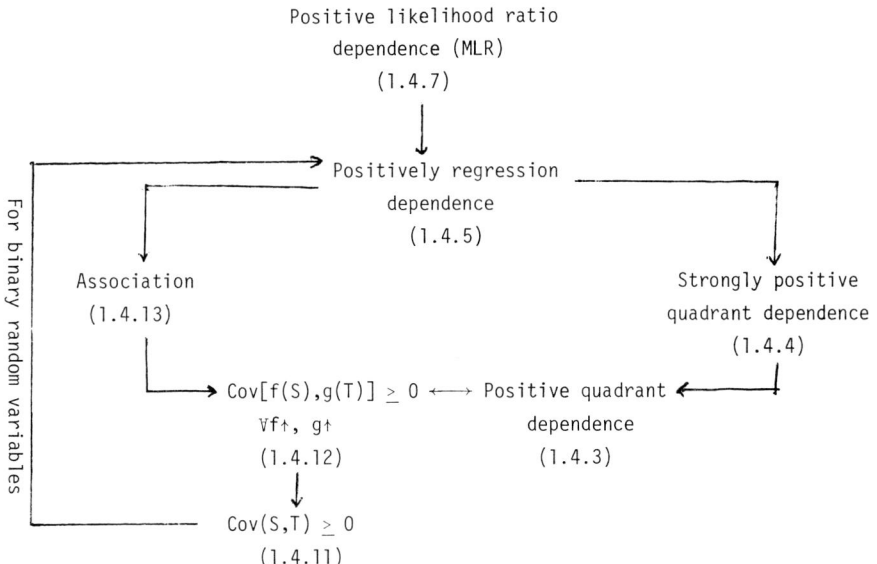

Figure 1.1

1.5. Majorization in Multivariate Distributions

Over the last fifty years, majorization and Schur functions have been applied to develop several useful inequalities in many branches of mathematics. Majorization is a partial ordering in $\mathbb{R}^n$. A Schur function is a function that is monotone with respect to this partial ordering. Many well known inequalities arising in probability and statistics are equivalent to saying that certain functions are Schur functions. The language and techniques of Schur functions often lead to more transparent and elegant proofs of currently known inequalities and to further generalizations. These techniques as applied to probability and statistics have become popular in recent times.

In n dimensions the vector $\underline{a}$ is said to be majorized by the vector $\underline{b}$ (written $\underline{a} \prec \underline{b}$) if, upon reordering components to achieve
$$a_1 \geq a_2 \geq \ldots \geq a_n, \quad b_1 \geq b_2 \geq \ldots \geq b_n,$$
it follows that
$$\sum_{i=1}^{k} a_i \leq \sum_{i=1}^{k} b_i, \quad i = 1, 2, \ldots, n-1, \quad \sum_{i=1}^{n} a_i = \sum_{i=1}^{n} b_i.$$
See, e.g., Hardy, Littlewood and Pólya (1934, p. 49). Functions φ for which $\underline{a} \prec \underline{b}$ implies $\varphi(\underline{a}) \leq \varphi(\underline{b})$ are said to be Schur-convex; if $\varphi(\underline{a}) \geq \varphi(\underline{b})$ they are called Schur-concave. Such functions are permutation-symmetric, i.e., invariant under permutations of components of the arguments. A necessary and sufficient condition that a permutation-symmetric differentiable function be Schur-concave is that

(1.5.1) $$\left(\frac{\partial \varphi(\underline{x})}{\partial x_i} - \frac{\partial \varphi(\underline{x})}{\partial x_j}\right)(x_i - x_j) \leq 0$$

for all $i \neq j$, (Schur (1923) and Ostrowski (1952)). The above inequality is reversed for Schur-convex functions.

It is easily verified that

$$(a_1, a_2, \ldots, a_n) \succ \left(\frac{1}{n} \Sigma a_i\right)(1, 1, \ldots, 1)$$

for all vectors $\underline{a}$. Thus, when Σa_i is fixed, a Schur-concave function achieves a maximum at the point where all components are equal. Another useful observation is that

$$(1, 0, \ldots, 0) \succ \frac{1}{2}(1, 1, 0, \ldots, 0) \succ \frac{1}{3}(1, 1, 1, 0, \ldots, 0)$$
$$\succ \ldots \succ \frac{1}{n}(1, 1, \ldots, 1).$$

1.5.1 **Theorem.** (Marshall and Olkin (1974)). Suppose that the random variables $X_1, X_2, \ldots, X_n$ have a joint density $f(\cdot)$, that is Schur-concave. If $A \subset R^n$ is a Lebesgue-measurable set which satisfies

(1.5.2) $$\underline{y} \in A \text{ and } \underline{x} \prec \underline{y} \Rightarrow \underline{x} \in A,$$

then

$$\int_{A+\underline{\theta}} f(\underline{x}) d\underline{x} = P\{\underline{X} \in A + \underline{\theta}\}$$

is a Schur-concave function of $\underline{\theta}$.

Proof. We must show that if $\underline{\theta} \prec \underline{\xi}$, then

(1.5.3) $$\int_{A+\underline{\theta}} f(\underline{x}) d\underline{x} \geq \int_{A+\underline{\xi}} f(\underline{x}) d\underline{x}.$$

Because f and A are permutation-symmetric,

$$\int_{A+\underline{\theta}} f(\underline{x}) d\underline{x} = \int_{A+\pi(\underline{\theta})} f(\underline{x}) d\underline{x}$$

for all permutations π. Thus, we can assume that

$$\theta_1 \geq \theta_2 \geq \ldots \geq \theta_n \text{ and } \xi_1 \geq \xi_2 \geq \ldots \geq \xi_n.$$

According to a result of Hardy, Littlewood and Pólya (1934), $\underline{\theta}$ can be derived from $\underline{\xi}$ by a finite number of pairwise averages of components. Consequently, we can assume that $\theta_i = \xi_i$ if $i \neq j$ or k, where $j < k$. Of course, $\theta_j + \theta_k = \xi_j + \xi_k = 2\delta$ say. If $\xi_j = \delta + \alpha$ and $\theta_j = \delta + \beta$, then $\xi_k = \delta - \alpha$, $\theta_k = \delta - \beta$ and $\alpha > \beta > 0$.

Let $u = x_j + x_k$ and $v = x_j - x_k$. To obtain (1.5.3), integrate first on v, conditionally with other variables held fixed. Observed that

$$\{v: (x_1, \ldots, x_{j-1}, \frac{u+v}{2}, x_{j+1}, \ldots, x_{k-1}, \frac{u-v}{2}, x_{k+1}, \ldots, x_k) \in A + \underline{\xi}\}$$
$$= \{v: (x_1 - \xi_1, \ldots, x_{j-1} - \xi_{j-1}, \frac{(u-2\delta)+(v-2\alpha)}{2}, x_{j+1} - \xi_{j+1}, \ldots, x_{k-1} - \xi_{k-1},$$

$$\frac{(u-2\delta)-(v-2\alpha)}{2}, x_{k+1}-\xi_{k+1},\ldots,x_n-\xi_n) \in A\} \equiv B_\alpha.$$

The set B_0 is symmetric and convex in v. Moreover, as a function of v,

$$f(\underline{x}) \equiv f(x_1,\ldots,x_{j-1}, \frac{u+v}{2}, x_{j+1},\ldots,x_{k-1}, \frac{u-v}{2}, x_{k+1},\ldots,x_n)$$

is symmetric and unimodal. Thus by the theorem of Anderson (1955) (see Lemma 1.3.2),

$$\int_{B_\beta} f(\underline{x})dv \geq \int_{B_\alpha} f(\underline{x})dv.$$

The inequality is preserved upon integrating out the remaining variables to yield (1.5.3).

1.5.2 <u>Definition</u>. Denumerably many random variables $X_1, X_2,\ldots$ are called exchangeable, if the joint distribution of any subset of m of them does not depend upon the fact that which ones are included in the set but only on m, the size of the set.

Note that the condition that f is Schur-concave implies that f is permutation-symmetric; this is just the condition that the random variables $X_1, X_2,\ldots,X_n$ are exchangeable.

It should be pointed out that the exchangeability for the finite case, $X_1,\ldots,X_n$ is as follows: The joint distribution of $(X_{\pi 1},\ldots,X_{\pi n})$ is the same as that of $(X_1,\ldots,X_n)$, for every permutation π of $\{1,2,\ldots,n\}$ (see Kingman (1978), Tong (1979)).

Condition (1.5.2) is satisfied whenever A is a permutation-symmetric convex set. The condition that f is Schur-concave is equivalent to the condition that for each constant c, $\{\underline{y}: f(\underline{y}) \geq c\}$ satisfies (1.5.2). Thus the condition is satisfied if $\{\underline{y}: f(\underline{y}) \geq c\}$ is convex (called unimodal by Anderson (1955)), and f is permutation If $f(\underline{x}) = \prod_{i=1}^{n} g(x_i)$, then f is Schur-concave if and only if log g is concave. Note that if log g is concave, g is called strongly unimodal. Since unimodality is a very strong property, it allows only a very limited set of useful conclusions to be drawn. In particular, it does not hold that the marginal distributions of a unimodal distribution are unimodal, nor that the convolutions of unimodal distributions are again unimodal. Barndorff-Nielson (1978) has discussed these useful properties and has also given some examples.

Now, we discuss Schur-convexity of the elliptically contoured densities introduced in Section 1.2.

1.5.3 <u>Theorem</u>. (Marshall and Olkin (1974)). If the joint density f has the form $f(\underline{x}) = g(\underline{x}'\Lambda\underline{x})$, where g is a (strictly) decreasing function and $\Lambda = (\lambda_{ij})$ is positive definite with $\lambda_{11} =\ldots= \lambda_{nn}$ and $\lambda_{ij} = \lambda$ when $i \neq j$, then f is Schur-concave.

<u>Proof</u>. We shall show that $f(\underline{x}) = g(\underline{x}'\Lambda\underline{x})$ is unimodal in the sense that sets of the form $\{\underline{x}: g(\underline{x}'\Lambda\underline{x}) \geq c\}$ are convex.

To check that $\{\underline{x}: g(\underline{x}'\Lambda\underline{x}) \geq c\}$ is convex, suppose that $g(\underline{y}'\Lambda\underline{y}) \geq c$ and $g(\underline{z}'\Lambda\underline{z}) \geq c$. Then $\underline{y}'\Lambda\underline{y} \leq g^{-1}(c)$ and $\underline{z}'\Lambda\underline{z} \leq g^{-1}(c)$. Because $\underline{x}'\Lambda\underline{x}$ is convex in $\underline{x}$; this implies

$$[\alpha\underline{y} + (1-\alpha)\underline{z}]' \Lambda [\alpha\underline{y} + (1-\alpha)\underline{z}] \leq g^{-1}(c)$$

and hence

$$g([\alpha\underline{y} + (1-\alpha)\underline{z}]' \Lambda [\alpha\underline{y} + (1-\alpha)\underline{z}]) \geq c.$$

Marshall and Olkin (1974) have discussed the following examples for various densities with Schur-concavity property.

1.5.4 **Example.** Multivariate normal distribution.

If $X_1, X_2, \ldots, X_n$ are exchangeable and are jointly normally distributed, then Theorem 1.5.3 shows that their joint density is Schur-concave.

1.5.5 **Example.** Multivariate "t" distribution.

If $U_1, \ldots, U_n$ are exchangeable and jointly normally distributed, and if Z^2 is distributed independently as a chi-square $(Z \geq 0)$, then $X_1 = \frac{U_1}{Z}, \ldots, X_n = \frac{U_n}{Z}$ have a joint density f of the form $g(\underline{x}'\Lambda\underline{x})$ where $g(w)$ is proportional to $(1+w)^{-a}$, $a > 0$. Thus Theorem 1.5.1 again applies to show that f is Schur-concave.

1.5.6 **Example.** Multivariate beta distribution.

If $U_1, \ldots, U_n$ and Z are independent random variables with a chi-square distribution and $U_1, \ldots, U_n$ are identically distributed, then from Olkin and Rubin (1964),

$$X_i = \frac{U_i}{\Sigma U_j + Z}, \quad i = 1, 2, \ldots, n$$

have a multivariate beta distribution with joint density of the form

(1.5.4) $$f(\underline{x}) = k \ (\pi x_i^{r-1}) \ (1-\Sigma x_j)^{s-1}.$$

Here, condition (1.5.1) is easily verified directly. Thus $f(\underline{x})$ given in (1.5.4) is Schur-concave.

1.5.7 **Example.** Multivariate chi-square distribution.

Suppose $S = (s_{ij})$ is the sample covariance matrix based on a sample size $N \geq p$ from a p-variate normal distribution with covariance matrices $\Sigma = (\sigma_{ij})$, $\sigma_{ii} = \sigma^2$, $\sigma_{ij} = \sigma^2 \rho$, $1 \leq i, j \leq p$, $i \neq j$. The joint density of $(s_{11}, \ldots, s_{pp})$ has the form of a mixture of independent chi-square densities (which are log concave). Consequently, the joint density of $s_{11}, \ldots, s_{pp}$ is Schur-concave.

1.5.8 Example. Multivariate "F" distribution.

A particular multivariate "F" distribution which arises in statistical contexts is generated as follows: $U_1,\ldots,U_n$ each has a chi-square distribution with $r \geq 2$ degrees of freedom, and Z has a chi-square distribution with s degrees of freedom and independent of $U_1,U_2,\ldots,U_n$. Since $U_1,\ldots,U_n$ and Z have log-concave densities

$$X_1 = \frac{U_1}{Z},\ldots,X_n = \frac{U_n}{Z}$$

have a joint density f which satisfies the conditions of Theorem 1.5.1. Moreover, the marginals of f are "F" densities.

1.5.9 Example. Multivariate gamma distribution.

If $U_1,\ldots,U_n$ and Z are independent, U_i's have a gamma distribution with density

$$f(u;\alpha) = \lambda(\lambda u)^{\alpha-1} e^{-\lambda u}/\Gamma(\alpha), \quad \alpha \geq 1,$$

and Z has a gamma distribution with density $f(u;\beta)$ then

$$X_1 = U_1+Z,\ldots,X_n = U_n+Z$$

are jointly distributed with marginal densities $f(u; \alpha+\beta)$.

1.5.10 Example. Non-central chi-square distribution.

This distribution function is given by

$$F_{\underline{\theta}}(t) = \frac{1}{(2\pi)^{n/2}} \int_{\{\Sigma x_i^2 \leq t\}} \exp[-\frac{1}{2}\Sigma(x_i-\theta_i)^2]d\underline{x}$$

$$= \frac{1}{(2\pi)^{n/2}} \int_{\{\underline{x} \in A-\underline{\theta}\}} \exp(-\frac{1}{2}\Sigma x_i^2)d\underline{x},$$

where $A = \{\underline{x}: \Sigma x_i^2 \leq t\}$.

Since A is convex and since the integrand is a product of log-concave densities, it follows from Theorem 1.5.1 that $F_{\underline{\theta}}$ is Schur-concave in $\underline{\theta}$. It is well known that $F_{\underline{\theta}}$ can be written as a mixture of central chi-square distribution functions where the mixing distribution is Poisson with parameter $\Sigma\theta_i^2$. One can show that $F_{\underline{\theta}}$ is decreasing in $\beta = \Sigma\theta_i^2$ either by direct differentiation or via total positivity. Since $\varphi(\underline{\theta}) = \Sigma\theta_i^2$ is Schur-convex, this implies that $F_{\underline{\theta}}$ is Schur-concave in $\underline{\theta}$.

1.5.11 Example. Non-central "t" distribution.

The distribution function of the non-central "t" is given by

$$F_{\underline{\theta}}(t) = \int_{\{\Sigma x_i \leq t\sqrt{s}\}} \exp[-\frac{1}{2}\Sigma(x_i-\theta_i)^2]s^{\frac{1}{2}n-1}e^{-\frac{1}{2}s}d\underline{x}ds.$$

The set $A_s = \{\underline{x}: \Sigma x_i \leq t\sqrt{s}\}$ are all convex in $\underline{x}$, and we can write

$$F_{\underline{\theta}}(t) = \int_{s\geq 0} \{\int_{\{\underline{x}\in A_s\}} \exp[-\frac{1}{2}\Sigma(x_i-\theta_i)^2]d\underline{x}\} s^{\frac{1}{2}n-1} e^{-\frac{1}{2}s} dx$$

$$= \int_{s\geq 0} \{\int_{\{\underline{x}\in A_s-\underline{\theta}\}} \exp[-\frac{1}{2}\Sigma x_i^2]d\underline{x}\} s^{\frac{1}{2}n-1} e^{-\frac{1}{2}s} dx.$$

The inner integral is Schur-concave in $\underline{\theta}$ for each fixed s by Theorem 1.5.1, and hence the mixture $F_{\underline{\theta}}(t)$ is Schur-concave in $\underline{\theta}$.

1.5.12 Example. Non-central "F" distribution.

The distribution function of the non-central "F" distribution can be written as

$$F_{\underline{\theta}}(t) = \int_{\{\Sigma x_i^2 \leq st\}} \exp[-\frac{1}{2}\Sigma(x_i-\theta_i)^2] s^{\frac{1}{2}n-1} e^{-\frac{1}{2}s} d\underline{x} ds$$

$$= \int_{s\geq 0} \{\int_{\underline{x}\in\{\Sigma x_i^2 \leq st\}} \exp[-\frac{1}{2}\Sigma(x_i-\theta_i)^2]d\underline{x}\} s^{\frac{1}{2}n-1} e^{-\frac{1}{2}s} ds.$$

Here the argument used for the non-central "t" requires little modification to show that the non-central "F" distribution function is Schur-concave in $\underline{\theta}$.

References

[1] Alam, K. (1973). On a multiple decision rule. Ann. Statist. 1, 750-755.

[2] Anderson, T. W. (1955). The integral of a symmetric convex set and some probability inequalities. Proc. Amer. Math. Soc. 6, 170-176.

[3] Barndorff-Nielson, O. (1978). Information and Exponential Families in Statistical Theory. John Wiley & Sons, New York.

[4] Bickel, P. J. (1965). Some contributions to the theory of order statistics. Proc. Berkeley Symp. Math. Statist. Prob., 1, 575-591. Univ. of California Press, Berkeley, California.

[5] Blomqvist, N. (1950). On a measure of dependence between any two random variables. Ann. Math. Statist. 21, 593-600.

[6] Bonnesen, T. and Fenchel, W. Theorie der Konvexen Käper, Chelsea Publishing Company, New York, 1948.

[7] Das, S. C. The numerical evaluation of a class of integrals II. Proc. Cambridge Philos. Soc. 52, 442-448, 1956.

[8] Das Gupta, S., Eaton, M. L., Olkin, I., Perlman, M., Savage, I. R. and Sobel, M. Inequalities on the probability content of convex regions for elliptically contoured distributions. Proc. Sixth Berkeley Symp. Math. Statist. Prob. 2, 241-265. Univ. of California Press, Berkeley, California, 1970.

[9] Doornbos, R. and Prins, H. J. On slippage tests. Indag. Math. 20, 38-55, 438-447, 1958.

[10] Dunnett, C. W. and Sobel, M. (1955). Approximation to the probability integral and certain percentage points of a multivariate analogue of Student's t-distribution. Biometrika 42, 258-260.

[11] Eaton, M. L. (1967). Some optimum properties of ranking procedures. Ann. Math. Statist. 38, 124-137.

[12] Efron, B. (1965). Increasing properties of Pólya frequency functions. Ann. Math. Statist. 36, 272-279.

[13] Esary, J. D., Proschan, F. and Walkup, D. W. (1967). Association of random variables, with applications. Ann. Math. Statist. 38, 1466-1474.

[14] Gupta, S. S. (1956). On a decision rule for a problem in ranking means. Inst. of Statist., Univ. of North Carolina, Mimeo. Ser. No. 150. (Tables AI and AII).

[15] Gupta, S. S. (1963). Probability integrals of multivariate normal and multivariate t. Ann. Math. Statist. 34, 792-828.

[16] Gupta, S. S. and Huang, D. Y. (1980). On an essentially complete class of multiple decision procedures. To appear in JSPI.

[17] Gupta, S. S., Nagel, K. and Panchapakesan, S. (1973). On the order statistics from equally correlated normal random variables. Biometrika 60, 403-413.

[18] Gupta, S. S. and Panchapakesan, S. (1972). On a class of subset selection procedures. Ann. Math. Statist. 43, 814-822.

[19] Gupta, S. S. and Hsu, J. C. (1977). On the monotonicity of Bayes subset selection procedures. Proceedings of the 41st Session of the International Statistical Institute, Vol. 47, Book 4, 208-211.

[20] Hardy, G. H., Littlewood, J. E. and Pólya, G. (1934). Inequalities. Cambridge Univ. Press, Cambridge, England.

[21] Hoel, D. G. (1970). On the monotonicity of the OC of an SPRT. Ann. Math. Statist. 41, 310-314.

[22] Hollander, M., Proschan, F. and Sethuraman, J. (1977). Functions decreasing in transposition and their applications in ranking problems. Ann. Statist. 5, 722-733.

[23] Hsu, J. C. (1977). On some decision-theoretic contributions to the problem of subset selection. Mimeo. Ser. #491, Dept. of Statist., Purdue Univ., W. Lafayette, Indiana.

[24] Ihm, P. (1959). Numerical evaluation of certain multivariate normal integrals. Sankhyā 21, 363-366.

[25] Jogdeo, K. (1970). A simple proof of an inequality for multivariate normal probabilities of rectangle. Ann. Math. Statist. 41, 1357-1359.

[26] Karlin, S. (1968). Total Positivity, Vol. I. Stanford Univ. Press, Stanford, California.

[27] Karlin, S. and Rubin, H. (1956). The theory of decision procedures for distributions with monotone likelihood ratio. Ann. Math. Statist. 27, 272-299.

[28] Kingman, J.F.C. (1978). Uses of exchangeability. Ann. Prob. 6, 183-197.

[29] Lehmann, E. L. (1955). Ordered families of distributions. Ann. Math. Statist. 26, 399-419.

[30] Lehmann, E. L. (1959). Testing Statistical Hypotheses. John Wiley.

[31] Lehmann, E. L. (1966). Some concepts of dependence. Ann. Math. Statist. 37, 1137-1153.

[32] Marshall, A. W. and Olkin, I. (1974). Majorization in multivariate distributions. Ann. Math. Statist. 2, 1180-1200.

[33] Marshall, A. W. and Olkin, I. (1979). Majorization and Schur Functions. Academic Press, New York.

[34] Moran, P.A.P. (1956). The numerical evaluation of a class of integrals. Proc. Cambridge Philos. Soc. 52, 230-233.

[35] Olkin, I. and Rubin, H. (1964). Multivariate beta distributions and independence properties of the Wishart distribution. Ann. Math. Statist. 35, 261-269.

[36] Oosterhoff, J. (1969). Combination of One-Sided Statistical Tests. Mathematical Centre Tracks 28, Amsterdam.

[37] Ostrowski, A. (1952). Sur quelque applications des functions convexes et concaves au sens de I. Schur. J. Math. Pure Appl. 31, 253-292.

[38] Patil, G. P. and Bosewell, M. T. (1970). A characteristic property of the multivariate normal density function and some of its applications. Ann. Math. Statist. 41, 1970-1977.

[39] Ruben, H. (1954). On the moments of order statistics in samples from normal populations. Biometrika 41, 200-227.

[40] Schur, I. (1923). Über eine Klasse von Mittelbildengen mit Anwendungen auf die Determinatentheorie. Sitzber Berl. Math. Ges. 22, 9-20.

[41] Sidák, Z. (1968). On multivariate normal probabilities of rectangles: Their dependence on correlations. Ann. Math. Statist. 39, 1425-1434.

[42] Slepian, D. (1962). The one-sided barrier problem for Gaussian noise. Bell System Tech. J. 41, 463-501.

[43] Stuart, A. (1958). Equally correlated variates and the multinomial integral. J. Roy. Statist. Soc. Ser. B 20, 373-378.

[44] Sherman, S. (1955). A theorem on convex sets with applications. Ann. Math. Statist. 26, 763-766.

[45] Tong, Y. L. (1980). Probability Inequalities in Multivariate Distributions. Academic Press, New York.

[46] Tukey, J. W. (1958). A problem of Berkson, and minimum variance orderly estimators. Ann. Math. Statist. 29, 588-592.

CHAPTER 2
MULTIPLE DECISION THEORY: A GENERAL APPROACH

In the theory and practice of statistical inference, multiple decision problems are encountered in many experimental situations. The classical methods for analyzing data do customarily employ hypothesis testing in most situations. In such cases, when the hypothesis is rejected, one wants to know in which of a number of possible ways the actual situation (true state of nature) differs from the one postulated by the null hypothesis. If, in the formulation of the problem, we consider only two decisions (reject or not reject the hypothesis), we will not only neglect to differentiate between certain alternative decisions but may also be using an inappropriate acceptance region for the hypothesis. Moreover, the traditional approach to hypotheses testing problems is not formulated in a way to answer the experimenter's question, namely, how to identify the "best" (in some sense) population? For example, the method of the least significant differences based on the t-test has been used in the past to detect differences between the true unknown means of different varieties and thereby choosing the population which is the "best", say, the one with the largest mean. But this method is indirect, less efficient and does not provide an overall probability of a correct decision. The remark (criticism) is also valid, to some extent, for methods based on multiple comparison techniques. The traditional approach does not allow for a decision if the null hypothesis is not rejected. Furthermore, when performing a test one may commit one of two errors: rejecting the hypothesis when it is true (error of the first kind) or accepting it when it is false (error of the second kind). It is desirable to carry out the test in a manner which keeps the probabilities of the two types of errors to a minimum. It is customary to assign a bound to the probability of incorrectly rejecting the hypothesis when it is true, and to attempt to minimize the other probability subject to this condition. Unfortunately, when the number of observations is given, both probabilities cannot be controlled simultaneously by the classical approach (see Lehmann (1959)). Kiefer (1977) gave an example (see Example 2.1 below) to show that for some sample values an appropriate test does not exhibit any detailed data-dependent measure of conclusiveness that conveys our stronger feeling in favor of the alternative hypothesis:

2.1 <u>Example</u>. Suppose we observe a normally distributed random variable X with mean θ and unit variance, and must decide between the two simple hypotheses H_0: $\theta = -1$ and H_1: $\theta = 1$. The test rejects H_0 if $X \geq 0$ and has the type I and II probabilities of errors $(\alpha, \beta) = (0.16, 0.16)$. Thus we make the same decision in favor of H_1 whether $X = 0.5$ or $X = 5$.

To enforce Kiefer's point, let us consider another example due to Schaafsma (1969). In this example (Example 2.2 below), Schaafsma's example points out that the Neyman-Pearson formulation is not always satisfactory and reasonable.

2.2 __Example__. Suppose an observation x is obtained from the normal $N(\mu,1)$ distribution in order to test the hypothesis H_0: $\mu = 0$ against the alternative H_1: $\mu = \mu_1$ where μ_1 is a given positive constant. In order to protect ourselves against the serious error of the first kind we use a small value of α, say, $\alpha = 0.05$. Then the Neyman-Pearson theory provides the test φ, where φ is the characteristic function of the critical region (u_α, ∞) where $u_\alpha = 1.645$. This result is not reasonable when $\mu_1 > 2u_\alpha$, since when an error of the first kind is regarded as much more serious than an error of the second kind then it is reasonable to require that

$$\alpha = E_{\mu=0}\{\varphi(X)\} < 1 - E_{\mu=\mu_1}\{\varphi(X)\} = \beta.$$

Hence α has, in some sense, to depend on μ_1 but the Neyman-Pearson theory does not take this into consideration.

In the preceding paragraphs, we have discussed various difficulties associated with the hypothesis testing formulation. Thus there arises the need for a modification of this theory and for alternative ways to attack such problems.

The approach in terms of Wald's decision theory (1950) provides an effective tool to overcome the above-mentioned difficulties in some reasonable ways. Actually, the problems of hypothesis testing can be formulated as general multiple decision problems. To this end, we first define that the space $\mathcal{A}$ of actions of the statistician consists of a finite number ($k \geq 2$) of elements, $\mathcal{A} = \{a_1, \ldots, a_k\}$. In practice, there are two distinct types of multiple decision problems. In one the parameter space Θ is partitioned into k subsets $\Theta_1, \ldots, \Theta_k$, according to the increasing value of a real-valued function $r(\underline{\theta})$, $\underline{\theta} \in \Theta$. The action a_i is preferred if $\underline{\theta} \in \Theta_i$. This type of multiple decision problem is called monotone. This approach has been studied by Karlin and Rubin (1956) and Brown, Cohen and Strawderman (1976). For example, in comparing two treatments with means θ_1 and θ_2, an experimenter might have only a finite number of actions available; among these the experimenter may have preference based on the magnitudes of the differences of the means $\theta_2 - \theta_1$: A particular case occurs when one may choose from the three alternatives:
 (i) prefer treatment 1 over treatment 2,
 (ii) prefer treatment 2 over treatment 1,
(iii) no preference, (cf. Ferguson (1967)).

Another important class of multiple decision problems arises in selection problems where the treatments are classified into a superior category (the selected items) and an inferior one. In general, selection problems have been treated under several different formulations. One basic distinction, pointed out by Lehmann (1961), corresponds to Model I and II cases in the analysis of variance. In Model I, the treatments being classified are considered fixed; only the observations made on each of them are random. In Model II, on the other hand, the treatments themselves are drawn at random from some population and would, therefore, change under replications of the experiment.

Lehmann (1961) (and earlier authors, see Paulson (1949), Bechhofer (1954), Seal

(1955), and Gupta (1956)) have formulated the selection problems according to the following goals:

Goal 1. We wish to select only a single population (if possible the best one): the variety to be planted, the production method we are going to adopt, etc. As a slight generalization, we may wish to select a fixed number, say, t, good populations.

Goal 2. The number of populations to be selected (or the subset size) is not fixed in advance but is determined by the observations. This arises, for example, when we wish to select all worthwhile treatments, or if we want to be reasonably sure that the selected group contains the best treatment.

Goal 3. The subset size is determined by the observed data subject to an upper bound specified in advance. It may, for example, be desirable to investigate all treatments that appear promising but budget restrictions may limit the research program to the investigation of at most three treatments.

Several authors had thought about these preceding goals; among the early investigators are Paulson (1949), Bahadur (1950), Bahadur and Robbins (1950). The formulation of such procedures in the framework of selection and ranking procedures has been generally accomplished either using the fixed subset size indifference zone approach or the (random-size) subset selection approach. The former approach was introduced by Bechhofer (1954). Substantial contributions to the early and subsequent developments in the subset selection theory have been made by Gupta starting with his work in 1956.

Indifference Zone Selection Approach

Bechhofer (1954) considered the problem of ranking k normal means. In order to explain the basic formulation, consider the problem of selecting the population with the largest mean from k normal populations with unknown means μ_i, $1 \leq i \leq k$, and a common known variance σ^2. Let $\bar{X}_i$, $i = 1,2,\ldots,k$, denote the means of k independent samples of size n from these populations. The "natural" procedure (which has been shown to have optimum properties by Bahadur and Goodman (1952), Hall (1959), Lehmann (1966), Eaton (1967) and Miescke (1979b)) is to select the population that yields the largest $\bar{X}_i$'s. The experimenter would, of course, need a guarantee that this procedure will pick the population having the largest μ_i with a probability not less than a specified level P*. Since we do not know the true configuration of the μ_i's, we look for the least favorable configuration (LFC) for which the probability of a correct selection (PCS) is minimized. This LFC is given by $\mu_1 = \ldots = \mu_k$; the corresponding PCS = $\frac{1}{k}$ and hence the probability guarantee cannot be met whatever be the sample size n. A natural modification is to insist on the minimum probability guarantee whenever the best population is sufficiently superior to the next best. In other words, the experimenter specifies a positive constant Δ^* and requires that the PCS is at least P^* whenever $\mu_{[k]} - \mu_{[k-1]} \geq \Delta^*$, where $\mu_{[1]} \leq \cdots \leq \mu_{[k]}$ denote the ordered means. Now the minimization of PCS is over the part Ω_{Δ^*} of the parameter space in which $\mu_{[k]} - \mu_{[k-1]} \geq \Delta^*$. The complement of Ω_{Δ^*} is called the indifference zone. The LFC in Ω_{Δ^*} is given by $\mu_{[1]} = \cdots = \mu_{[k-1]} = \mu_{[k]} - \Delta^*$. The problem is then to determine the minimum sample size required in order to have PCS $\geq P^*$ for all $\underline{\mu}$ such that $\mu_{[k]} - \mu_{[k-1]} \geq \Delta^*$.

Bechhofer's formulation (1954) is more general than what is described above. His general ranking problem includes, for example, selection of the t(≥ 2) best populations.

Subset Selection Approach

In the subset selection approach, the goal is to select a nonempty subset of the populations so as to include the best population. Here the size of the selected subset is random and is determined by the observations themselves. In the case of normal populations with unknown means $\mu_1,\ldots,\mu_k$, and a common known variance σ^2, the rule proposed by Gupta (1956) selects the population that yields $\bar{X}_i$ if and only if

$$\bar{X}_i \geq \max_{1 \leq j \leq k} \bar{X}_j - \frac{d\sigma}{\sqrt{n}},$$

where $d = d(k,P^*) > 0$ is determined so that the PCS is at least P^*. Here a correct selection is the selection of any subset that includes the population with the largest μ_i. Thus, the LFC is considered with regard to the whole parameter space Ω. Under this formulation, for given k and P^* we determine the constant d. The rule explicitly involves n. In general, the rule will involve a constant which depends on k, P^* and n. The performance of a subset selection procedure is studied by evaluating the expected subset size depending on the parameters or its supremum over the parameter space Ω. Various ways to study the performance of subset selection procedures, are to consider the measure of loss related to an incorrect selection ICS($\underline{\theta}$,a) where $\underline{\theta} = (\theta_1,\ldots,\theta_k)$, and the number of elements $|a|$ in the selected subset a. More recently, Goel and Rubin (1977) studied the subset selection problem from a Bayesian point of view using loss functions that are linear combinations of $\theta_{[k]} - \max_{j \in a} \theta_j$ and $|a|$. Miescke (1979b) considered a similar problem using linear loss functions. Bickel and Yahav (1977) studied the behaviour of Bayes procedures as $k \to \infty$ using loss functions that are linear combinations of ICS($\underline{\theta}$,a) and $\theta_{[k]} - \sum_{j \in a} \theta_j/|a|$. Chernoff and Yahav (1977), employing Monte Carlo techniques, compared the integrated risks with respect to exchangeable normal priors of Bayes procedures, and Bechhofer type procedures using loss functions that are linear combinations of $\theta_{[k]} - \max_{j \in a} \theta_j$ and $\theta_{[k]} - \sum_{j \in a} \theta_j/|a|$. Gupta and Hsu (1978) used Monte Carlo study which parallels that of Chernoff and Yahav, in that exchangeable normal priors are used but differs in that the loss functions considered are linear combinations of ICS($\underline{\theta}$,a) and $|a|$. Several classical methods have been compared by Gupta and Hsu (1978).

There are some deficiencies in all these approaches. The subset selection approach is conservative in that we do not use the information we might have about the parameters. On the other hand, the indifference zone approach is not efficient enough since we may not know the real situation, i.e., the indifference zone, in most practical problems. In view of these deficiencies, several authors (see, for example, Desu (1970)) have tried to modify the formulation described above (for more appropriate reference see Gupta and Panchapakesan (1979)). The indifference zone and subset selection approaches can, for example, be combined to guarantee the infimum of the probability of a correct selection over a preference zone and minimize the supremum of the expected size of the selected subset over the parameter space. We restrict the class of decision rules to the rules which guarantee the probability requirement as above. Among this class we might choose an optimal rule evaluated by the expected size of the selected subset (see Gupta and Huang (1980a)). In many cases, we do not know whether the true parametric configuration belongs to the preference zone. If the best and the second best are not very much different, it is reasonable to select a subset. We would like to keep the size of the subset under control. This can be achieved by increasing the sample size which also increases the probability of a correct selection.

After the selection procedure has been applied, one can, roughly speaking, say that all populations (treatments) in the selected set are "homogeneous". In particular, if all populations are selected, one can use it as a rough (quick) test of the homogeneity hypothesis.

Moreover, consider the action space $G_0 = \{a_1,\ldots,a_k\}$ consisting of k elements corresponding to k populations. The action a_i is to select the ith population for the problem of selecting (identifying) the best population. This is the case with Bechhofer's fixed subset size procedures under the indifference zone formulation (see Bechhofer (1954)). On the other hand, let G be the action space consisting of the 2^k subsets of the set $\{1,2,\ldots,k\}$. This is the so-called (random size) subset selection rule. Since $G_0 \subset G$, the associated decision rules of G_0 is a subclass $\mathcal{D}_0$ of the class $\mathcal{D}$ associated with G. Thus it is more reasonable to obtain optimal rules in $\mathcal{D}$ than in $\mathcal{D}_0$.

A related problem for selection in which the number of populations selected does not exceed a given number m (< k), was solved by Gupta and Santner (1973) and Santner (1975).

It should be pointed out that more recently, Kiefer (1975, 1976, 1977) and Brownie and Kiefer (1977) have made several significant contributions in the very interesting area of multiple decision problems based on conditional inference.

We now mention in passing the nature of the multiple comparison problems. These techniques are designed for making inferences concerning all pairwise differences or contrasts. The main contributions to this field are due to Tukey and Scheffé.

Miller (1966) has summarized the developments in the theory and methods of multiple comparisons to study the appropriate alternatives in hypotheses testing problems. Further developments in multiple comparisons during the years 1966-1976 have been reviewed in a later paper by Miller (1977). Some recent developments in the field of multiple decision procedures based on tests are also interesting. Here the works of Gabriel (1969), Holm (1977), Marcus, Peritz and Gabriel (1976), Shaffer (1978), and Lehmann and Shaffer (1979) - based on the idea of stagewise multiple test procedures - are relevant (see Chapter 6 for more details).

Prior Information

For the information concerning the a priori distribution of a parameter, many statisticians feel that except in rare situations it is likely to be incomplete. Hence the use of a Bayes rule on some systematically produced choice for completely known a priori distribution, as advocated by the Bayesian school, is difficult to justify. This appears to be the case sometimes even if the a priori distribution is known fairly accurately. Robbins (1964) has suggested that attention should be paid to the case in which it is known only that the distribution of the parameter is a member of some given family Γ of distributions. Blum and Rosenblatt (1967) have used this idea in statistical inference. Thus if in a problem it may be reasonable to assume the existence of an a priori distribution function F, but it is unreasonable to assume perfect knowledge of F, the optimum procedure is the so-called Γ-minimax procedure which minimizes the supremum of the Bayes risks over a class Γ of priors. Some contributions to multiple decision problems using this criterion have been made by Randles and Hollander (1971), Gupta and Huang (1977) and Miescke (1979a). It should be noted that it is still necessary to use completely known prior information to show some optimality properties. We note also that if Γ has only one element, say F, the Γ-minimax rule coincides with the Bayes rule δ_F, while if Γ includes the set of all degenerate distributions over Θ then the Γ-minimax rule coincide with the minimax rule.

The situations occuring in practice with regard to any prior knowledge about a parameter θ usually lie between the two extremes just described; to this extent at least both principles of minimax and Bayes are often subject to criticism. One possible way out of this dilemma was mentioned already, namely, the Γ-minimax principle. Another possible way is to restrict our attention to a class of decision rules which are minimax but satisfy some other conditions which may be of interest and concern to the experimenter. This is the so-called restricted minimax approach.

Essential Completeness Nature of Decision Rules

In statistical decision theory, the concept of admissible or complete classes of strategies is generally thought of as providing the most satisfactory solution. whereas it may be difficult to say what to do in a statistical decision problem, it is easier to say what not to do, so that the statistician separates out from considerations all inadmissible strategies and presents the practical man with what is at least an essentially complete class of procedures (Gupta and Huang (1980b) have studied this in the context of some multiple decision problems). The idea of essential completeness is to show that the choice of one from among these admissible procedures is then left to the best judgement, intuition, and past experience of the practical man. If the class is a small one, we have then achieved everything one can ask for, and the actual choice will then easily be made. The difficulty, however, is that the classes are usually much too big to be of real help. An important topic, then, is to study how to restrict a class of rules to be useful in order to investigate reasonable rules with some good properties.

Furthermore, most research works in this area lay emphasis on completely randomized designs. However, in many practical situations one may require the blocking of treatments in order to cut down on bias and improve the precision of the experiment. Recently, Bechhofer and Tamhane (1979) have considered some multiple comparison problems in this framework. It is important to continue research in multiple decision problems which have a bearing on the preceding aspect.

Multiple decision procedures have been applied to some problems in nonparametric statistical methods, regression analysis, analysis of variance and reliability theory, among others. Much more research along these lines need to be accomplished. Derivation of suitable statistical methods for the data obtained from a proposed experimental design is a very important subject to be investigated. The consideration of robustness of multiple decision problems is also a widely open topic. The study of multistage procedures has received some attention during the last few years. Optimality of such procedures needs to be further investigated.

References

[1] Bahadur, R. R. (1950). On a problem in the theory of k populations. Ann. Math. Statist. 21, 362-375.

[2] Bahadur, R. R. and Goodman, L. A. (1952). Impartial decision rules and sufficient statistics. Ann. Math. Statist. 23, 553-562.

[3] Bahadur, R. R. and Robbins, H. (1950). The problem of the greater mean. Ann. Math. Statist. 21, 469-487; Correction: 22(1951), 310.

[4] Bechhofer, R. E. (1954). A single sample multiple decision procedure for ranking means of normal populations with known variances. Ann. Math. Statist. 25, 16-39.

[5] Bechhofer, R. E. and Tamhane, A. C. (1979). Incomplete block designs for comparing treatments with a control I II III. Technique Report No. 414, School of Operation Research and Industrial Engineering, Cornell Univ., Ithaca, New York.

[6] Bickel, P. J. and Yahav, J. A. (1977). On selecting a set of good populations. Statistical Decision Theory and Related Topics II, (ed. S. S. Gupta and D. S. Moore), 37-55. Academic Press.

[7] Blum, J. R. and Rosenblatt, J. (1967). On partial a priori information in statistical inference. Ann. Math. Statist. 38, 1671-1678.

[8] Brown, L. D., Cohen, A. and Strawderman, W. E. (1976). A complete class theorem for strict monotone likelihood ratio with applications. Ann. Statist. 4, 712-722.

[9] Brownie, C. and Kiefer, J. (1977). The ideas of conditional confidence in the simplest setting. Commun. Statist.-Theor. Meth., A6 (8), 691-751.

[10] Chernoff, H. and Yahav, J. A. (1977). A subset selection problem employing a new criterion. Statistical Decision Theory and Related Topics II, (ed. S. S. Gupta and D. S. Moore), 93-119. Academic Press.

[11] Desu, M. M. (1970). A selection problem. Ann. Math. Statist. 41, 1596-1603.

[12] Eaton, M. L. (1967). Some optimum properties of ranking procedures. Ann. Math. Statist. 38, 124-137.

[13] Ferguson, T. S. (1967). Mathematical Statistics: A Decision Theoretic Approach. Academic Press, New York.

[14] Gabriel, K. R. (1969). Simultaneous test procedures-some theory of multiple comparisons. Ann. Math. Statist. 40, 224-250.

[15] Goel, P. K. and Rubin, H. (1977). On selecting a subset containing the best population - a Bayesian approach. Ann. Statist. 5, 969-983.

[16] Gupta, S. S. (1956). On a decision rule for a problem in ranking means. Ph.D. thesis, Univ. of North Carolina.

[17] Gupta, S. S. and Hsu, J. C. (1978). On the performance of some subset selection procedures. Comm. Statist. - Simula. Computation, B7(6), 561-591.

[18] Gupta, S. S. and Huang, D. Y. (1977). On some Γ-minimax selection and multiple comparison procedures. Statistical Decision Theory and Related Topics II (ed. S. S. Gupta and D. S. Moore), 139-155.

[19] Gupta, S. S. and Huang, D. Y. (1980a). A note on optimal subset selection procedure. To appear in Ann. Statist.

[20] Gupta, S. S. and Huang, D. Y. (1980b). An essentially complete class of multiple decision procedures. To appear in Journal of Statistical Planning and Inference.

[21] Gupta, S. S. and Panchapakesan, S. (1979). Multiple Decision Procedures. John Wiley & Sons, New York.

[22] Gupta, S. S. and Santner, T. J. (1973). On selection and ranking procedures - a restricted subset selection rule. Proceedings of the 39th Session of the International Statistical Institute, Vol. 45, Book 1, 409-417.

[23] Hall, W. J. (1959). The most economical character of Bechhofer and Sobel decision rules. Ann. Math. Statist. 30, 964-969.

[24] Holm, S. A. (1977). Sequentially rejective multiple test procedures. Statistical research report 1977-1, Univ. of Umea, Sweden.

[25] Karlin, S. and Rubin, H. (1956). The theory of decision procedures for distributions with monotone likelihood ratio. Ann. Math. Statist. 27, 272-299.

[26] Kiefer, J. (1975). Conditional confidence approach in multidecision problems. Proc. 4th Dayton Multivariate Conf. ed. P. R. Krishnaiah, Amsterdam: North Holland Publishing Co., 143-158.

[27] Kiefer, J. (1976). Admissibility of conditional confidence procedures. Ann. Math. Statist. 4, 836-865.

[28] Kiefer, J. (1977). Conditional confidence statements and confidence estimators. (With comments.) JASA 72, 789-827.

[29] Lehmann, E. L. (1957). A theory of some multiple decision problems I II. Ann. Math. Statist. 28, 1-25, 547-572.

[30] Lehmann, E. L. (1959). Testing Statistical Hypotheses. John Wily, New York.

[31] Lehmann, E. L. (1961). Some model I problems of selection. Ann. Math. Statist. 32, 990-1012.

[32] Lehmann, E. L. (1966). On a theorem of Bahadur and Goodman. Ann. Math. Statist. 37, 1-6.

[33] Marcus, R., Peritz, E. and Gabriel, K. R. (1976). On closed testing procedures with special reference to ordered analysis of variance. Biometrika 63, 655-660.

[34] Miescke, K. J. (1979a). Γ-minimax selection procedures in simultaneous testing problems. Mimeo. Ser. 79-1, Dept. of Statist., Purdue Univ., IN 47907.

[35] Miescke, K. J. (1979b). Bayesian subset selection for additive and linear loss functions. Commun. Statist. - Theor. Meth. A8(12), 1205-1226.

[36] Miller, R. G., Jr. (1966). Simultaneous Statistical Inference. New York, McGraw Hill Book Co.

[37] Miller, R. G., Jr. (1977). Developments in multiple comparisons 1966-1976. JASA 72, 779-788.

[38] Paulson, E. (1949). A multiple decision procedure for certain problems in analysis of variance. Ann. Math. Statist. 20, 95-98.

[39] Randles, R. H. and Hollander, M. (1971). Γ-minimax selection procedures in treatments versus control problems. Ann. Math. Statist. 42, 330-341.

[40] Robbins, H. (1964). The empirical Bayes approach to statistical decision problems. Ann. Math. Statist. 35, 1-20.

[41] Santner, T. (1975). A restricted subset selection approach to ranking and selection problems. Ann. Statist. 3, 334-349.

[42] Schaafsma, W. (1969). Minimax risk and unbiasedness for multiple decision problems of type I. Ann. Math. Statist. 40, 1684-1720.

[43] Shaffer, J. P. (1978). Control of directional errors with stagewise multiple test procedures. Report, Univ. of California, Berkeley.

[44] Seal, K. C. (1955). On a class of decision procedures for ranking means of normal populations. Ann. Math. Statist. 26, 387-398.

[45] Wald, A. (1950). Statistical Decision Functions. John Wiley & Sons, New York.

CHAPTER 3
MODIFIED MINIMAX DECISION PROCEDURES

3.1. Introduction

In this chapter, we shall discuss the modified principles of minimax criteria (such as Γ-minimax and restricted minimax) for multiple decision problems.

In Section 3.2, we study some optimal procedure for selecting good populations with respect to a control. Selecting the best population will be discussed in Section 3.3. Two different types of monotone rules are considered. Some classes of essentially complete or complete rules are studied in Section 3.4. We also formulate hypotheses testing as a selection problem. Corresponding to the errors of traditional approach, we define several kinds of errors. This is a very important concept to treat testing hypotheses.

3.2. The Problem of Selecting Good Populations With Respect to a Control

Let $\pi_1, \pi_2, \ldots, \pi_k$ be $k (\geq 2)$ independent populations. We wish to select a subset containing good populations. The quality of the ith population is characterized by a real-valued parameter θ_i, and a population is said to be

(3.2.1)   positive (or good) if $\theta_i \geq \theta_0 + \Delta$,

(3.2.2)   negative (or bad) if $\theta_i \leq \theta_0$,

where Δ is a given positive constant and θ_0 is either a given number or an unknown parameter.

As in the Neyman-Pearson theory of hypothesis-testing, there are two possible sources of error in any selections. There is the possibility of false positives, that is, populations which are selected although they are negative, and of false negatives, that is, populations which are not selected although they are positive. We shall focus attention on true positives, that is, on those positive populations which are included in the selected group. This is analogous to the replacement of the consideration of an error of the second kind by that of power in the Neyman-Pearson theory.

Our object is to define selection procedures to seek out the true positives while holding false positives to a minimum. It is very important to evaluate the performance of any proposed decision procedure. Reasonable criteria for the performance of a rule can be given in various ways.

For such problems, Lehmann (1961) proposes some criteria to measure how well a procedure to carry out its task of identifying the positive populations as follows:
(a) The expected number of true positives.
(b) The expected proportion of true positives, that is, the quantity (a) divided by the total number of positives.

These criteria are appropriate if it is desired to include in the selected group as many of the positive populations as possible.

(c) The probability of at least one true positive.

(d) The probability of including in the selected group the best population (that is, the population with the largest θ-value), provided it is positive.

(e) The probability of including all good populations.

As a measure of the performance of a procedure with respect to false positives we shall take either

(i) the expected number of false postives or

(ii) the expected proportion of false positives, that is, the quantity (i) divided by the total number of negatives.

As a generic notation for any one of the quantities (a) to (e), all of which depend on the parameter point $\underline{\theta}$ and the particular selection procedure δ under investigation, we shall use $S(\underline{\theta},\delta)$. Here it is understood that S is defined only for the set Ω' of those parameter points for which at least one of the populations is positive.

Similarly, we shall let $R(\underline{\theta},\delta)$ denote the quantity (i) or (ii). With these definitions of R and S, it is desirable to have $S(\underline{\theta},\delta)$ as large and $R(\underline{\theta},\delta)$ as small as possible.

We shall restrict the decision rules in a class C such that for any $\delta \in C$,

(3.2.3) $$\inf_{\underline{\theta}\in\Omega'} S(\underline{\theta},\delta) \geq \gamma,$$

then we wish to choose a $\delta_0 \in C$ satisfying

(3.2.4) $$\inf_{\delta\in C} \sup_{\underline{\theta}\in\Omega} R(\underline{\theta},\delta) = \sup_{\underline{\theta}\in\Omega} R(\underline{\theta},\delta_0),$$

where Ω denotes the whole parameter space.

The formulation in the preceding part, is designed to obtain a minimax decision rule subject to a side condition. Lehmann (1961) uses a result which is an immediate extension of the standard method of characterizing minimax solutions as Bayes solutions corresponding to a least favorable a priori distribution. He proves the following theorem:

3.2.1 <u>Theorem</u>. Let the probability density of $\underline{X}$ be denoted by p_0 when $\theta_1 = ... = \theta_k = \theta_0$, and by p_i when $\theta_i = \theta_0 + \Delta$ and the parameters θ_j for $j \neq i$ have a common value $\theta' < \theta_0 + \Delta$ determined so that the conditions below are satisfied. Suppose that $p_i(\underline{x})/p_0(\underline{x})$ is a nondecreasing function of a real-valued statistic T_i, that the distribution of T_i depends only on θ_i, is stochastically increasing in θ_i, and is independent of i. Then the procedure δ_0 satisfying (3.2.3) and (3.2.4) with S equal to any one of the quantities (a)-(d) and R defined by (i) as before, is given by

(3.2.5) $$\psi_i = 1, \lambda_0, 0, \text{ as } T_i >, =, < c,$$

where λ_0 and c are determined by

(3.2.6) $$E_{\theta_0+\Delta}\psi_i = \gamma.$$

Lehmann (1961) points out that this theorem provides the basis for determining the

sample size necessary to control the risks R and S at any desired levels. Suppose that we wish the selection procedure to satisfy
$$R(\underline{\theta},\delta) \leq \gamma' \quad \text{for all} \quad \underline{\theta} \in \Omega,$$
and
$$S(\underline{\theta},\delta) \geq \gamma \quad \text{for all} \quad \underline{\theta} \in \Omega'.$$

Then for the smallest sample size which constitutes a solution to this problem, the associated procedure δ_0 minimizes sup $R(\underline{\theta},\delta)$ subject to (3.2.3). If the conditions of the theorem are satisfied, δ_0 is given by (3.2.5) and hence satisfies (if we assume for simplicity that it is nonrandomized)
$$\sup R(\underline{\theta},\delta_0) = R(\underline{\theta}_0,\delta_0) = kP_{\theta_0}(T_i \geq c)$$
if R is given by (i). It further satisfies the condition
$$\inf S(\underline{\theta},\delta_0) = P_{\theta_0+\Delta}(T_i \geq c).$$

If we let $\gamma^* = \frac{\gamma}{k}$ when R is given by (i), the requires sample size is determined by the conditions
$$P_{\theta_0}(T_i \geq c) \leq \gamma^* \quad \text{and} \quad P_{\theta_0+\Delta}(T_i \geq c) \geq \gamma.$$

These are exactly the conditions appropriate for testing the hypothesis $\theta_i = \theta_0$ against the alternative $\theta_i = \theta_0+\Delta$ if we wish to have significance level γ^* and power at least γ.

Lehmann (1961) also discusses a slight generalization of Theorem 3.2.1 to the case of unequal samples. Some applications to the selection of variances and coefficients of variation are also studied by him.

Suppose we have some information relevant to the problem which may be based on past experience then it is important that we use this prior information. In such cases, procedures based on such prior information have been studied in the context of minimax solutions. Such procedures, the so-called Γ-minimax procedures, have been studied by Randles and Hollander (1971) for the problem of selecting populations which have larger translation parameters than that of a control population. We discuss this problem in the following part.

Let $X_0,\ldots,X_k$ be k+1 independent random observable variables with probability density function $f_0(x-\theta_0)$, $f_1(x-\theta_1),\ldots,f_k(x-\theta_k)$, respectively. The variables X_0, $X_1,\ldots,X_k$ may represent sufficient statistics from the control and k treatment populations, respectively. We assume that each $f_i(x)$, $i = 0,1,\ldots,k$, is a Pólya function of order two (PF$_2$), that is, if $x_1 < x_2$ and $y_1 < y_2$ then
$$\begin{vmatrix} f_i(x_1-y_1) & f_i(x_1-y_2) \\ f_i(x_2-y_1) & f_i(x_2-y_2) \end{vmatrix} \geq 0.$$
Hence $f_i(x-\theta_i)$ has monotone likelihood ratio property (MLR).

We shall discuss some selection procedures for selecting all positive populations while rejecting all negative ones.

Let L_1 denote the loss incurred whenever we fail to select a positive population and L_2, the loss for each negative population selected. If $\underline{X} = (X_0, X_1, \ldots, X_k)$, consider decision rules of the form

(3.2.7) $$\psi(\underline{x}) = (\psi_1(\underline{x}), \ldots, \psi_k(\underline{x})),$$

where $\psi_i(\underline{x})$ denotes the conditional probability of selecting the ith population given $\underline{X} = \underline{x}$. The loss function is then

(3.2.8) $$L(\underline{\theta}, \psi) = \sum_{i=1}^{k} L^{(i)}(\underline{\theta}, \psi_i),$$

where $\underline{\theta} = (\theta_0, \theta_1, \ldots, \theta_k)$ and

$$L^{(i)}(\underline{\theta}, \psi_i) = \begin{cases} L_1(1-\psi_i), & \text{if } \theta_i \geq \theta_0 + \Delta, \\ L_2\psi_i, & \text{if } \theta_i \leq \theta_0, \\ 0, & \text{otherwise.} \end{cases}$$

The risk function is

$$R(\underline{\theta}, \psi) = \sum_{i=1}^{k} R^{(i)}(\underline{\theta}, \psi_i)$$

where

$$R^{(i)}(\underline{\theta}, \psi_i) = \int_{\mathcal{S}_0} \int_{\mathcal{S}_1} \cdots \int_{\mathcal{S}_k} L^{(i)}(\underline{\theta}, \psi_i(\underline{x})) \prod_{i=0}^{k} [f_i(x_i - \theta_i) dx_i]$$

and $\mathcal{S}_i$ denotes the sample space of the random variable X_i. Thus

$$R(\underline{\theta}, \psi) = L_1 N_1 + L_2 N_2$$

where N_1 (N_2) is the expected number of positive (negative) populations rejected (selected). Note that there is no loss of generality in considering decision rules of the form given in (3.2.7). For any decision rule there exists a rule in the same class (3.2.7) with the same risk function. If $\tau(\underline{\theta})$ is a distribution over Θ then the expected risk of a procedure ψ is

$$\gamma(\tau, \psi) = \sum_{i=1}^{k} \gamma^{(i)}(\tau, \psi_i)$$

where

$$\gamma^{(i)}(\tau, \psi_i) = \int_{\Theta} R^{(i)}(\underline{\theta}, \psi_i) d\tau(\underline{\theta}).$$

Assume that partial prior information is available in the selection problem. Define $\Theta_P(i) = \{\underline{\theta} | \theta_i \geq \theta_0 + \Delta\}$ and $\Theta_N(i) = \{\underline{\theta} | \theta_i \leq \theta_0\}$ and assume that we are able to specify $\pi_i = P[\underline{\theta} \in \Theta_P(i)]$ and $\pi_i' = P[\underline{\theta} \in \Theta_N(i)]$ such that $\pi_i + \pi_i' \leq 1$ for $i = 1, 2, \ldots, k$. Define

$$\Gamma = \{\tau(\underline{\theta}) | \int_{\Theta_P(i)} d\tau(\underline{\theta}) = \pi_i \text{ and } \int_{\Theta_N(i)} d\tau(\underline{\theta}) = \pi_i', \text{ for } i = 1, 2, \ldots, k\}.$$

Consider the ith component problem, that is, the selection or rejection of the ith population when the parameter θ_0 is known. Note that X_0 is of no interest in this case. The loss function for the component problem is $L^{(i)}(\theta,\psi_i)$. Randles and Hollander (1971) have proved that the Γ-minimax decision rule, ψ^Γ, is of the form:

$$\psi_i^\Gamma = \begin{cases} 1, & \text{if } X_i \geq d_i \\ 0, & \text{if } X_i < d_i \end{cases}$$

for $i = 1,\ldots,k$, where each d_i is determined by the following

(3.2.9) $\quad L_2\pi_i' f_i(x-\theta_0) - L_1\pi_i f_i(x-\theta_0-\Delta) \leq, > 0$

as $x \geq, < d_i$.

Randles and Hollander (1971) give an example to illustrate the application of the above results as follows: Consider $k+1$ normal populations $N(\theta_i,\sigma^2)$, $i = 0,1,\ldots,k$, with θ_0 and σ^2 known. A random sample of size n_i is taken from each of the k populations $N(\theta_i,\sigma^2)$, $i = 1,\ldots,k$. In this case, d_i is given by

$$d_i = \theta_0 + \frac{\Delta}{2} + \sigma^2(\Delta n_i)^{-1} \ell n(L_2\pi_i'/L_1\pi_i).$$

Thus the Γ-minimax decision rule selects the ith population if and only if

$$\bar{X}_i \geq d_i$$

where $\bar{X}_i$ is the sample mean of the ith population.

In case of θ_0 unknown, Randles and Hollander (1971) obtain the following result: A Γ-minimax decision rule, ψ^Γ, is of the form

$$\psi_i^\Gamma = \begin{cases} 1, & \text{if } X_i - X_0 \geq d_i', \\ 0, & \text{if } X_i - X_0 < d_i', \end{cases}$$

where d_i' is determined by the following

$$L_2\pi_i' g_i(r) - L_1\pi_i g_i(r-\Delta) \leq, > 0$$

as $r \geq, < d_i'$ and

$$g_i(r) = \int_{-\infty}^{\infty} f_i(r+u) f_0(u) du,$$

for $i = 1,2,\ldots,k$.

Note that the proof of Randles and Hollander needs some modification in the case where θ_0 is unknown (see Miescke (1979)).

Randles and Hollander (1971) apply the above result to the following example. A sample of size n_i is taken from each of $k+1$, $N(\theta_i,\sigma^2)$ populations with σ^2 known. We confine attention to rules ψ where ψ_i is based on $\bar{X}_0$ and $\bar{X}_i$. Then a Γ-minimax decision rule selects the ith population if and only if

(3.2.11) $\bar{X}_i - \bar{X}_0 \geq \frac{\Delta}{2} + \frac{\sigma^2}{\Delta}(\frac{1}{n_i} + \frac{1}{n_0})\ln(L_2\pi_i'/L_1\pi_i).$

After discussing Γ-minimax decision rules, we shall now compare a Γ-minimax rule with a Bayes rule to shed some light on the robustness of the rules.

If a priori considerations yield a class of prior distributions over Θ, one method of utilizing such information is to select a member of the class and use the corresponding Bayes decision rule. Another approach is to find a rule which is Γ-minimax with respect to the class of priors given. Thus Bayes rules corresponding to prior distributions in Γ are natural competitors for a Γ-minimax rule. Randles and Hollander (1971) have made a comparison between the Bayes procedure which is based on the specific prior information of independent normal distributions with the Γ-minimax procedure which is based on the less stringent prior specifications of π_i and π_i' for i = 1,2,...,k. Assume that $L_1 = L_2 = 1$ so that the expected risk becomes the expected number of wrong decisions where a wrong decision is defined to be the rejection of a positive population or the selection of a negative one. One meaningful comparison is found by examining the increase in expected risk which results from the use of the Γ-minimax procedure when $\underline{\theta}$ is distributed over Θ according to $\tau^*(\underline{\theta}) = \prod_{i=0}^{k} \tau_i(\theta_i)$, where $\tau_i(\theta_i)$ is the a priori distribution of θ_i which is normal with mean α_i and variance γ_i^2, for i = 0,1,...,k. Randles and Hollander (1971) have shown that the Γ-minimax procedure is seen to have only slightly higher expected risk under independent normal priors in the cases given.

3.3 On the Problem of Selecting the Best Population

Let $\pi_1, \pi_2,...,\pi_k$ represent $k(\geq 2)$ independent populations (treatments) and let $X_{i1},...,X_{in_i}$ be n_i independent random observations from π_i. the quality of the ith population π_i is characterized by a real-valued parameter θ_i, usually unknown. Let $\Omega = \{\underline{\theta}|\underline{\theta}' = (\theta_1,...,\theta_k)\}$ denote the parameter space. In practice, the experimenter wishes to select the "best" population. We may define how to measure the bestness in various ways. The concept of $\tau_{ij} = \tau_{ij}(\underline{\theta})$ is introduced to measure the separation between π_i and π_j. We assume that there exists a monotonically nonincreasing function h such that $\tau_{ji} = h(\tau_{ij})$. Let $\tau_i = \min_{j \neq i} \tau_{ij}$, $1 \leq i \leq k$. We define $\tau^* = \max_{1 \leq \ell \leq k} \tau_\ell$. The population associated with τ^* will be called the best population. Note that it is possible to have more than one best population. For defining the bestness, we partition the space Ω into the corresponding sets. Let

$$\Omega_i = \{\underline{\theta}|\tau_{ij} \geq \tau_0, \forall j \neq i\}, 1 \leq i \leq k,$$

and

$$\Omega_0 = \Omega - \bar{\Omega}, \bar{\Omega} \neq \phi,$$

where $\bar{\Omega} = \bigcup_{i=1}^{k} \Omega_i$, and where τ_0 and τ_{ii} are assumed known with $\tau_0 > \tau_{ii}$ for i = 1,...,k.

It should be noted that if $\underline{\theta} \in \Omega_i$, then $\tau_i \geq \tau_j$ for all j, since for some j, $j \neq i$, $\tau_{ji} = h(\tau_{ij}) \leq h(\tau_0) \leq h(\tau_{ii}) = \tau_{ii} < \tau_0$. Thus if $\underline{\theta} \in \Omega_i$, then π_i is the best population. A selection of a subset containing the best population is called a correct selection (CS).

To illustrate the above notation, we assume that independent observations are drawn from π_i, which has a normal distribution with unknown mean θ_i ($i = 1,\ldots,k$) and known variance σ^2. We define $\tau_{ij} = \theta_i - \theta_j$; then $\tau_i = \theta_i - \theta_{[k]}$ if $\theta_i < \theta_{[k]}$ and $\tau_i = \theta_i - \theta_{[k-1]}$ if $\theta_i = \theta_{[k]}$, where $\theta_{[1]} \leq \ldots \leq \theta_{[k]}$. In this case, $\tau_{ii} = 0$ for all i and the population with the largest mean $\theta_{[k]}$, is the best. If, instead, $\tau_{ij} = \theta_j - \theta_i$ then the population with the smallest mean $\theta_{[1]}$, would be the best. In the above example, $h(t) = -t$, which is a decreasing function.

Let the observed sample vector be denoted by $\underline{X}' = (\underline{X}_1',\ldots,\underline{X}_k')$ where $\underline{X}_i$ has components $X_{i1},\ldots,X_{in_i}$, $i = 1,\ldots,k$. Let $\delta = (\delta_1,\ldots,\delta_k)$ be a selection procedure where $\delta_i(\underline{x})$ is the probability of selecting π_i ($1 \leq i \leq k$) based on the observed vector $\underline{X} = \underline{x}$. As measures of goodness of a selection rule, consider two quantities $R(\underline{\theta},\delta)$ and $S(\underline{\theta},\delta)$. We define $S(\underline{\theta},\delta) = P_{\underline{\theta}}(CS|\delta)$ and $R(\underline{\theta},\delta) = \sum_{i=1}^{k} R^{(i)}(\underline{\theta},\delta_i)$, where $R^{(i)}(\underline{\theta},\delta_i) = P\{\text{selecting } \pi_i|\delta\}$. Thus, here $R(\underline{\theta},\delta)$ is the expected size of the selected subset. For a specified γ, ($0 < \gamma < 1$), we restrict attention to the class C of all δ such that

(3.3.1) $$S(\underline{\theta},\delta) \geq \gamma \text{ for } \underline{\theta} \in \bar{\Omega}.$$

We are interested in constructing an optimal procedure δ^0 in C which minimizes the supremum of $R(\underline{\theta},\delta)$ over Ω for all $\delta \in C$, i.e.,

(3.3.2) $$\sup_{\underline{\theta} \in \Omega} R(\underline{\theta},\delta^0) = \min_{\delta \in C} \sup_{\underline{\theta} \in \Omega} R(\underline{\theta},\delta).$$

We restrict attention to those selection procedures which depend on the observations only through a sufficient statistic for $\underline{\theta}$.

Let the statistic Z_{ij} be based on the n_i and n_j independent observations from π_i and π_j ($i,j = 1,2,\ldots,k$), respectively, and suppose that for any i, the statistic $\underline{Z}_i' = (Z_{i1},\ldots,Z_{ik})$ is invariant sufficient under a transformation group G and let $\underline{\tau}_i' = (\tau_{i1},\ldots,\tau_{ik})$ be a maximal invariant under the induced group $\bar{G}$. It is well known (see Ferguson (1967)) that the distribution of $\underline{Z}_i$ depends only on $\underline{\tau}_i$. For any i, let the joint density of Z_{ij}, $\forall j \neq i$, be $p_{\underline{\theta}}(\underline{z}_i)$ with SIP. Let $p_{\underline{\theta}}(\underline{z}_i)$ be denoted by $p_0(\underline{z}_i)$ when $\tau_{i1} = \ldots = \tau_{ik} = \tau_{ii} = c$ (constant) and by $p_i(\underline{z}_i)$ when $\tau_{i1} = \ldots = \tau_{ik} = \tau_0$, $1 \leq i \leq k$.

In the normal means example, a choice of Z_{ij} might be $\bar{X}_i - \bar{X}_j$, where $\bar{X}_i = \frac{1}{n_i} \sum_{\ell=1}^{n_i} X_{i\ell}$ and $\bar{X}_j = \frac{1}{n_j} \sum_{\ell=1}^{n_j} X_{j\ell}$. Let ν be a σ-finite measure on $\mathbb{R}^{k-1}$.

Gupta and Huang (1980a) provide a solution to the restricted minimax problem as stated in (3.3.1) and (3.3.2).

3.3.1 **Theorem.** Suppose that for any i, $p_i(\underline{z}_i)/p_0(\underline{z}_i)$ is nondecreasing in $\underline{z}_i$ and assume that $p_{\underline{\theta}}(\underline{z})$ has stochastically increasing property. If $R(\underline{\theta}, \delta^0)$ is maximized at $\tau_{ij} = \tau_{ii} = $ constant, for all i,j, where δ^0 is given by

$$\delta_i^0(\underline{z}_i) = \begin{cases} 1 & \text{if } p_i(\underline{z}_i) > cp_0(\underline{z}_i), \\ \lambda_i & \text{if } p_i(\underline{z}_i) = cp_0(\underline{z}_i), \\ 0 & \text{if } p_i(\underline{z}_i) < cp_0(\underline{z}_i), \end{cases}$$

such that c (> 0) and λ_i are determined by $\int \delta_i^0 p_i = \gamma$, $1 \leq i \leq k$. Then $\delta^0 = (\delta_1^0, \ldots, \delta_k^0)$ minimizes $\sup_{\underline{\theta} \in \Omega} R(\underline{\theta}, \delta)$ subject to $\inf_{\underline{\theta} \in \bar{\Omega}} S(\underline{\theta}, \delta) \geq \gamma$. Then δ^0 is the restricted minimax rule for this problem.

As an application of Theorem 3.3.1, consider the following example:

3.3.2 **Example.** Let $\pi_1, \pi_2, \ldots, \pi_k$ be k independent normal populations with means $\theta_1, \ldots, \theta_k$ and common known variance $\sigma^2 = 1$. The ordered θ_i's are denoted by $\theta_{[1]} \leq \cdots \leq \theta_{[k]}$. It is assumed that there is no prior knowledge of the correct pairing of the ordered and the unordered θ_i's. Our goal is to select a nonempty subset of the k populations so as to include the population associated with $\theta_{[k]}$.

Let $\bar{X}_i$, $1 \leq i \leq k$, denote the sample means of independent samples of size n from these populations. The likelihood function of $\underline{\theta}$ is then

$$p_{\underline{\theta}}(\underline{x}) = \prod_{i=1}^{k} p_{\theta_i}(\bar{x}_i),$$

where

$$p_{\theta_i}(\bar{x}_i) = \frac{n^{\frac{1}{2}}}{(2\pi)^{\frac{1}{2}}} e^{-\frac{n}{2}(\bar{x}_i - \theta_i)^2}, \quad 1 \leq i \leq k.$$

Let $\tau_{ij} = \tau_{ij}(\underline{\theta}) = \theta_i - \theta_j$, $1 \leq i, j \leq k$, $\tau_0 = \Delta > 0$, $\bar{\Omega} = \{\underline{\theta} | \theta_{[k]} - \theta_{[k-1]} \geq \Delta\}$ and $Z_{ij} = \bar{X}_i - \bar{X}_j$, $1 \leq i, j \leq k$. Let $\underline{z}'_i = (z_{i1}, \ldots, z_{ik})$, $\underline{\tau}'_i = (\tau_{i1}, \ldots, \tau_{ik})$, then since $Z_{ii} = 0$ and $\tau_{ii} = 0$, $\forall i$, the joint density of Z_{ij}, $j = 1, \ldots, k$, $j \neq i$, is given by

$$p_{\underline{\theta}}(\underline{z}_i) = (2\pi)^{-\frac{k-1}{2}} |\Sigma|^{-\frac{1}{2}} \exp\{-\frac{1}{2}(\underline{z}_i - \underline{\tau}_i)' \Sigma^{-1} (\underline{z}_i - \underline{\tau}_i)\},$$

$$\Sigma_{(k-1) \times (k-1)} = \frac{1}{n} \begin{bmatrix} 2 & & 1 \\ & \ddots & \\ 1 & & 2 \end{bmatrix}$$

is the covariance matrix of Z_{ij}'s. Since

$$\frac{p_i(\underline{z}_i)}{p_0(\underline{z}_i)} = \exp \frac{1}{2} \{\underline{z}'_i \Sigma^{-1} \underline{\Delta} + \underline{\Delta}' \Sigma^{-1} \underline{z}_i - \underline{\Delta}' \Sigma^{-1} \underline{\Delta}\} = \exp\{\frac{n\Delta}{2k}(z_{i1} + \ldots + z_{ik})\}$$

is nondecreasing in z_{ij}, $j \neq i$, where $\underline{\Delta}' = (\Delta,\ldots,\Delta)$. Hence

$$\frac{p_i(\underline{z}_i)}{p_0(\underline{z}_i)} > c$$

is equivalent to

$$\bar{x}_i > \frac{1}{k-1} \sum_{j \neq i} \bar{x}_j + d.$$

Since $R(\underline{\theta},\delta^0) = \sum_{i=1}^{k} P\{\bar{X}_i > \frac{1}{k-1} \sum_{j \neq i} \bar{X}_j + d\}$ is the expected size of the selected subset for Seal's average-type procedure δ^0 (1955), the following result of Berger (1977) and Bjørnstad (1980) applies

$$\sup_{\theta \in \Omega} R(\underline{\theta},\delta^0) = R(\underline{0},\delta^0) \text{ iff } \inf_{\theta \in \Omega} S(\underline{\theta},\delta^0) \geq \frac{k-1}{k}.$$

Since the right hand side is equivalent to

$$\Phi(((k-1)/k)^{\frac{1}{2}} n^{\frac{1}{2}} d) \leq \frac{1}{k},$$

the left hand side for every fixed $\Delta > 0$ holds if and only if

$$\gamma = 1 - \Phi((\frac{k-1}{k})^{\frac{1}{2}} n^{\frac{1}{2}} (d-\Delta)) \geq 1 - \Phi(\Phi^{-1}(\frac{1}{k}) - (\frac{k-1}{k})^{\frac{1}{2}} n^{\frac{1}{2}} \Delta),$$

where $\Phi(\cdot)$ is the cdf of the standard normal. Therefore, if for any $\Delta > 0$, γ is chosen in such a way that the preceding inequality holds, then the result of the Theorem 3.3.1 can be applied.

Another approach to restrict to a class of reasonable procedures has been studied by Bjørnstad (1980) as Schur-procedures as follows.

We assume that $X_1,\ldots,X_k$ are independent and X_i has continuous density $f(x-\theta_i)$, $1 \leq i \leq k$. Let $X_{[1]} \leq \ldots \leq X_{[k-1]}$ be the ordered $\{X_j: j \neq i\}$. We define a subset selection procedure by:

(3.3.3) $$\psi(\underline{x}) = (\psi_1(\underline{x}),\ldots,\psi_k(\underline{x})),$$

where $\psi_i(\underline{x}) = P\{\text{selecting } \pi_i | \underline{X} = \underline{x}\}$, $i = 1,\ldots,k$.

We shall require that at least one population is selected which implies that

(3.4) $$\sum_{i=1}^{k} \psi_i(\underline{x}) \geq 1 \quad \forall \underline{x}.$$

3.3.3 <u>Definition</u> (see Nagel (1970)). ψ is said to be just if $\psi_i(\underline{x})$ is nondecreasing in x_i and non-increasing in each x_j, $j \neq i$, for $i = 1,\ldots,k$.

Since we are dealing with a permutation-symmetric location parameter case it is natural that a selection procedure is also translation-invariant and permutation invariant. In order to define precisely a permutation-invariant procedure, let π be a permutation of $(1,\ldots,k)$ such that πi is the new position of element i, i.e., $\pi(1,\ldots,k) = (\pi^{-1}1,\ldots,\pi^{-1}k)$. Then the permutation $\pi\underline{x}$ of $\underline{x} \in \mathbb{R}^k$ is defined by

$$(\pi\underline{x})_i = x_{\pi^{-1}i}, \text{ for } i = 1,\ldots,k.$$

3.3.4 **Definition.** $\psi = (\psi_1,\ldots,\psi_k)$ is permutation-invariant if $\psi_i(\underline{x}) = \psi_{\pi i}(\pi\underline{x})$ for $i = 1,\ldots,k$ and all $(\pi,\underline{x})$.

Let η be the class of just, permutation-invariant and translation-invariant procedures. For $\underline{u} \in \mathbb{R}^k$ define

(3.3.5)
$$\underline{u}_i^* = (u_1-u_i,\ldots,u_{i-1}-u_i, u_{i+1}-u_i,\ldots,u_k-u_i).$$

3.3.5 **Definition** (Bjørnstad (1980)). A subset selection procedure $\psi = (\psi_1,\ldots,\psi_k)$ is said to be a Schur-procedure if $\psi \in \eta$ and there is a Schur-concave function ψ' such that $\psi_i(\underline{x}) = \psi'(\underline{x}_i^*)$, for $i = 1,\ldots,k$.

Let
$$\underline{c} = (c_1,\ldots,c_{k-1}) \in \mathbb{R}^{k-1},\ c_i \geq 0 \text{ and } \sum_{i=1}^{k-1} c_i = 1.$$

The procedure ψ^c is defined by

(3.3.6)
$$\psi_i^c = 1 \text{ iff } \sum_{j=1}^{k-1} c_j X_{[j]} - X_i \leq d(\underline{c}).$$

Here $d(\underline{c})$ is determined such that $\inf_{\theta \in \Omega} P_\theta(CS|\psi) = \gamma$.

Bjørnstad (1980) has proved that the class of procedures ψ^c described above is the same as the following

$$C = \{\psi^c : \sum_{i=1}^{k} c_i = 1 \text{ and } \gamma \geq \frac{1}{k}\}.$$

Hence we have the following result.

3.3.6 **Theorem** (Bjørnstad (1980)). Let
$$C_0 = \{\psi^c \in C : c_1 \leq \ldots \leq c_{k-1}\}.$$
Assume that $\psi^c \in C$. Then

$\psi^c \in C_0$ iff ψ^c is a Schur-procedure.

Note that the two procedures ψ^a and ψ^m defined as below are both in C_0 and are therefore Schur-procedures:

(1) Seal procedure (1955):
$$\psi_i^a(\underline{x}) = 1 \text{ iff } x_i \geq \frac{1}{k-1} \sum_{j \neq i} x_j - c,$$

(2) Gupta procedure (1965):
$$\psi_i^m(\underline{x}) = 1 \text{ iff } x_i \geq \max_{1 \leq j \leq k} x_j - d.$$

Suppose we have partial information about $\underline{\theta}$. Define the loss function for $\delta = (\delta_1,\ldots,\delta_k)$ as follows:

$$L(\underline{\theta},\delta) = \sum_{i=1}^{k}\sum_{j=1}^{k} L^{(i)}(\underline{\theta},\delta_j)$$

where, for any i, $1 \le i \le k$, and $\underline{\theta} \in \Omega_i$, $L^{(r)}(\underline{\theta},\delta_j) = 0$ for all $r \ne i$,

$$L^{(i)}(\underline{\theta},\delta_j) = \begin{cases} w_{ij}\delta_j & \text{for } j \ne i, \\ 0 & \text{for } j = i, \end{cases}$$

and w_{ij} ($j \ne i$) are given positive numbers and $w_{ii} = 0$. For $\underline{\theta} \in \Omega_0$, the loss is zero.

If $\rho(\underline{\theta})$ is a distribution over Ω, then the risk of a procedure is

$$\gamma(\rho,\delta) = \int_{\Omega}\int_{\mathcal{X}} L(\underline{\theta},\delta)f_{\underline{\theta}}(\underline{x})d\mu(\underline{x})d\rho(\underline{\theta})$$

$$= \sum_{i,j=1}^{k} \int_{\Omega_i}\int_{\mathcal{X}} L^{(i)}(\underline{\theta},\delta_j)f_{\underline{\theta}}(\underline{x})d\mu(\underline{x})d\rho(\underline{\theta})$$

$$= \sum_{i=1}^{k} \int_{\Omega_i}\int_{\mathcal{X}} \sum_{j=1}^{k} w_{ij}\delta_j(\underline{x})f_{\underline{\theta}}(\underline{x})d\mu(\underline{x})d\rho(\underline{\theta}),$$

where μ is the Lebesgue measure on $\mathbb{R}^k$. We specify $p_i = P[\underline{\theta} \in \Omega_i]$, $\sum_{i=1}^{k} p_i \le 1$. Define

$$\Gamma = \{\rho(\underline{\theta}) | \int_{\Omega_i} d\rho(\underline{\theta}) = p_i, i = 1,2,\ldots,k\}.$$

Gupta and Huang (1977) have obtained Γ-minimax rules as follows.

3.3.7 <u>Theorem</u>. For any i, $1 \le i \le k$, if there is a $\underline{\theta}_i^* \in \Omega_i$ such that

(3.3.7)
$$\sup_{\underline{\theta}\in\Omega_i} \int_{\mathcal{X}} \sum_{j=1}^{k} w_{ij}\delta_j^0(\underline{x})f_{\underline{\theta}}(\underline{x})d\mu(\underline{x})$$

$$= \int_{\mathcal{X}} \sum_{j=1}^{k} w_{ij}\delta_j^0(\underline{x})f_{\underline{\theta}_i^*}(\underline{x})d\mu(\underline{x}),$$

where

$$\delta_i^0(\underline{x}) = \begin{cases} 1, & \text{if } L_i(\underline{x}) < \min_{j \ne i} L_j(\underline{x}), \\ a_i, & = \quad , \\ 0, & > \quad , \end{cases}$$

$L_i(\underline{x}) = \sum_{j=1}^{k} w_{ji}p_j f_{\underline{\theta}_j^*}(\underline{x})$, $1 \le i \le k$, and $\sum_{1}^{k} a_i = 1$. Then $\delta^0 = (\delta_1^0,\ldots,\delta_k^0)$ is a Γ-minimax rule.

Existence of $\underline{\theta}_{-i}^*$ for the condition (3.3.7).

Let $\pi_1, \pi_2, \ldots, \pi_k$ be k independent populations with densities $f_{\theta_1}(\cdot), f_{\theta_2}(\cdot), \ldots, f_{\theta_k}(\cdot)$, respectively. Let $f_{\theta_i}(x) = f(x-\theta_i)$ have MLR, $1 \leq i \leq k$. Let $\tau_{ij} = \theta_i - \theta_j$ and $\tau_0 = \Delta$, $\Delta > 0$. Then $\Omega_i = \{\underline{\theta} | \theta_i \geq \max_{j \neq i} \theta_j + \Delta\}$, $1 \leq i \leq k$.

For any i, $1 \leq i \leq k$, we assume that $w_{ij} = w_i$, $\forall j \neq i$, $1 \leq i \leq k$. Since $\sum_{j=1}^{k} \delta_j^0 = 1$ and $w_{ii} = 0$, hence $\sum_{j=1}^{k} w_{ij} \delta_j^0(\underline{x}) f_{\underline{\theta}}(\underline{x}) = w_i(1-\delta_i^0(\underline{x})) f_{\underline{\theta}}(\underline{x})$. Thus the condition (3.3.7) is satisfied if

(3.3.8) $$\inf_{\underline{\theta} \in \Omega_i} E_{\underline{\theta}} \delta_i^0(\underline{X}) = E_{\underline{\theta}_{-i}^*} \delta_i^0(\underline{X}),$$

where $\underline{\theta}_{-i}^* = (\theta_0^*, \ldots, \theta_0^*, \theta_0^* + \Delta, \theta_0^*, \ldots, \theta_0^*)$, θ_0^* fixed.

In order to prove (3.3.8), we proceed as follows:
$$L_i(\underline{x}) < \min_{j \neq i} L_j(\underline{x})$$

is equivalent to

(3.3.9) $$w_j \pi_j \frac{f_{\theta_0^* + \Delta}(x_j)}{f_{\theta_0^*}(x_j)} < w_i \pi_i \frac{f_{\theta_0^* + \Delta}(x_i)}{f_{\theta_0^*}(x_i)}, \quad \forall j \neq i.$$

Hence $\delta_i^0(\underline{x})$ is nondecreasing in x_i and nonincreasing in x_j, for all $j \neq i$. Therefore,

$$\inf_{\underline{\theta} \in \Omega_i} E_{\underline{\theta}} \delta_i^0(\underline{X}) = \inf_{\substack{\theta_j = \theta^*, j \neq i \\ \theta_i = \theta^* + \Delta}} E_{\underline{\theta}} \delta_i^0(\underline{X}).$$

To determine θ^* in the normal case, let

$$f_{\underline{\theta}}(\underline{x}) = \prod_{j=1}^{k} \frac{\sqrt{n}}{\sqrt{2\pi}} e^{-\frac{n}{2}(\bar{x}_j - \theta_j)^2}.$$

For any i, $1 \leq i \leq k$, the condition (3.3.9) is equivalent to

$$\max_{j \neq i} \left[\bar{x}_j + \frac{1}{n\Delta} \log \frac{w_j \pi_j}{w_i \pi_i} \right] < \bar{x}_i.$$

Since for $\underline{\theta} \in \{\underline{\theta} | \theta_j = \theta^*, j \neq i, \theta_i = \theta^* + \Delta\}$,

$$E_{\underline{\theta}} \delta_i^0(\underline{X}) = \int_{-\infty}^{\infty} \prod_{j \neq i} \Phi(x + \Delta - \frac{1}{n\Delta} \log \frac{w_j \pi_j}{w_i \pi_i}) d\Phi(x)$$

is independent of $\underline{\theta}$, we can choose any fixed θ^*. Note that Gupta and Huang (1977) (1977) discuss this problem in the more general case of (3.3.9).

3.4 Essentially Complete Classes of Decision Procedures

Let $\pi_1,\ldots,\pi_k$ represent $k (\geq 2)$ independent populations and let $X_{i1},\ldots,X_{in_i}$ be n_i independent random observations from π_i. The quality of the ith population π_i is characterized by a real-valued parameter θ_i, usually unknown. Let $\Omega = \{\underline{\theta}|\underline{\theta}' = (\theta_1,\ldots,\theta_k)\}$ denote the parameter space. Let $\tau_{ij} = \tau_{ij}(\underline{\theta})$ be a measure of separation between π_i and π_j. We assume that there exists a monotone nonincreasing function h such that $\tau_{ji} = h(\tau_{ij})$. Let $\tau_i = \min_{j \neq i} \tau_{ij}$ and $\Omega_i = \{\underline{\theta}|\tau_i \geq \tau_{ii}\}$, $1 \leq i \leq k$. In the sequel we assume τ_{ii} to be known, $1 \leq i \leq k$, and moreover, that $\tau_{ij} \geq \tau_{ii}$, $\forall j \neq i$, in case all θ's are equal. We assume further that $\Omega = \bigcup_{i=1}^{k} \Omega_i$ holds. We call a population π_i the best population if $\underline{\theta} \in \Omega_i$, $i = 1,\ldots,k$. A selection of a subset containing the best population is called a correct selection (CS). It should be pointed out that in this case of ties the "best population" is considered to be a "tagged" one.

We will restrict attention to those selection procedures which depend upon the observations only through a sufficient statistic for $\underline{\theta}$. For $i \neq j$, let the statistic Z_{ij} be based on the n_i and n_j observations from π_i and π_j $(i = 1,2,\ldots,k)$, respectively, and suppose that for any i, $\underline{Z}'_i = (Z_{i1},\ldots,Z_{ik}) \in R^{k-1}$ is invariant sufficient under a transformation group G and let $\underline{\tau}'_i = (\tau_{i1},\ldots,\tau_{ik}) \in R^{k-1}$ be a maximal invariant under the induced group $\bar{G}$. It is well known that the distribution of $\underline{Z}_i$ depends only on $\underline{\tau}_i$. For any i, let the joint density of Z_{ij}, $\forall j \neq i$, be $p_{\underline{\tau}_i}(\underline{z}_i)$ be denoted by $p_i(\underline{z}_i)$ when $\tau_{i1} = \ldots = \tau_{ik} = \tau_{ii}$, where $\underline{z}'_i = (z_{i1},\ldots,z_{ik})$, $1 \leq i \leq k$. For any i, and fixed $z_{i\ell}$, $\ell = 1,\ldots,k$, $\ell \neq i, j$, and σ-finite measures μ on R' with product $\nu = \mu \times \ldots \times \mu$ on R^{k-1}, let

$$F_i^j(x) = \int_{-\infty}^{x} p_i(\underline{z}_i) d\mu(z_{ij}) \text{ and } F_i^{0j}(y) \in \{z|F_i^j(z) = y\}.$$

Note that $F_i^j(x)$ is a function of x and $z_{i\ell}$, $\ell = 1,\ldots,k$, $\ell \neq i,j$.

Let $\delta = (\delta_1,\ldots,\delta_k)$ be a selection procedure where $\delta_i(\underline{z})$ denotes the conditional probability of selecting π_i, $1 \leq i \leq k$, having observed $\underline{z}$. Let

$$S(\underline{\theta},\delta) = P(CS|\delta) \text{ and } R(\underline{\theta},\delta) = \sum_{i=1}^{k} R^i(\underline{\theta},\delta_i),$$

where

$$R^i(\underline{\theta},\delta_i) = \int \delta_i(\underline{z}_i) p_{\underline{\tau}_i}(\underline{z}_i) d\nu(\underline{z}_i), \; 1 \leq i \leq k.$$

Here $S(\underline{\theta},\delta)$ is the probability of correct selection and $R(\underline{\theta},\delta)$ is the expected value of the size (number of populations) of the selected subset. Our goal is to derive a selection procedure δ which controls $R(\underline{\theta},\delta)$ at points of the form $\tau_{i1} = \tau_{i2} = \ldots = \tau_{ik}$, $i = 1,2,\ldots,k$, and maximizes $\inf_{\underline{\theta} \in \Omega} S(\underline{\theta},\delta)$.

For given $Z_{i\ell}$, $\ell = 1,\ldots,k$, $\ell \neq i,j$, let

$$R_j^i(\delta_i) = \int \delta_i(\underline{z}_i)p_i(\underline{z}_i)d\mu(z_{ij}), \quad 1 \leq i, j \leq k.$$

A decision rule $\delta^{(1)} = (\delta_1^{(1)},\ldots,\delta_k^{(1)})$ is said to be "as good as" $\delta^{(2)} = (\delta_1^{(2)},\ldots,\delta_k^{(2)})$, if

$$\inf_{\underline{\theta}\in\Omega} S(\underline{\theta},\delta^{(1)}) \geq \inf_{\underline{\theta}\in\Omega} S(\underline{\theta},\delta^{(2)})$$

provided that

$$\int \delta_j^{(i)}(\underline{z}_j)p_j(\underline{z}_j)d\nu(\underline{z}_j) = \gamma_j, \quad 1 \leq j \leq k, \; i = 1,2,$$

where γ_j, $(0 < \gamma_j < 1)$, $1 \leq j \leq k$, are specified numbers.

We formulate the selection problem described above as hypotheses testing as follows:

$$H_0: \tau_{i1} = \ldots = \tau_{ik} = \tau_{ii} = c, \; H_{1j}: \tau_j > c, \; j = 1,\ldots,k,$$

for any i. The number γ_j is called the error of i-th kind while H_0 is true. In this case, we explain the expected size of selected subset as the total sum of all i-th kind of errors. The probability of correct selection can be interpreted as the power of the selection procedure.

Let $\mathcal{C}$ be the class of all decision rules δ such that

$$\int \delta_i(\underline{z}_i)p_i(\underline{z}_i)d\nu(\underline{z}_i) = \gamma_i, \quad 1 \leq i \leq k.$$

3.4.1 Definition. A probability density $f_{\underline{\theta}}(\underline{x})$ has generalized monotone likelihood ratio (GMLR) in $\underline{x}$, if for every i and for all fixed x_j, $j = 1,\ldots,k$, $j \neq i$, $f_{\underline{\theta}_1}(\underline{x})/f_{\underline{\theta}_2}(\underline{x})$ is nondecreasing in x_i, where $\underline{\theta}_\ell = (\theta_{\ell 1},\ldots,\theta_{\ell k})$, $\ell = 1,2$; $\theta_{1j} = \theta_{2j}$ for all $j \neq i$ and $\theta_{1i} > \theta_{2i}$.

Note that for $f_{\underline{\theta}}(\underline{x})$ with GMLR, if we fix all x_j's except x_i, then $f_{\underline{\theta}}(\underline{x})$ has TP_2 in x_i.

The following lemma is due to Karlin and Rubin (1956).

3.4.2 Lemma. If $p(x|w)$ is positive integrable and TP_2 and φ changes sign at most once in one-dimensional Euclidean space R', then

$$\psi(w) = \int p(x|w)\varphi(x)d\mu(x)$$

changes sign at most once. Moreover, if $\psi(w_0) = 0$, while for φ_1 and φ_2, the positive and negative part of φ (i.e., $\varphi = \varphi_1 - \varphi_2$, $\varphi_1, \varphi_2 \geq 0$),

$$\psi_i(w_0) = \int p(x|w_0)\varphi_i(x)d\mu(x) > 0, \; i = 1,2,$$

then $\psi(w) \geq 0$ for $w \geq w_0$ and $\psi(w) \leq 0$ for $w \leq w_0$.

3.4.3 Remark. It is useful to note that ψ changes sign in the same direction as φ, if it changes sign at all.

Now, we define a "monotone" selection rule as follows.

3.4.4 Definition. A selection rule $\delta = (\delta_1, \ldots, \delta_k)$ is called monotone if for any i and j, for fixed $z_{i\ell}$, $\ell = 1, \ldots, k$, $\ell \neq i,j$, $\delta_i(\underline{z}_i)$ is nondecreasing in z_{ij}.

Consider the class $\mathcal{C}$ defined earlier and the concept of completeness in the sense of the measure of "as good as" defined earlier. Gupta and Huang (1980b) have proved the following result.

3.4.5 Theorem. Let $p_{\underline{\tau}}(\underline{z})$ be continuous and let $p_{\underline{\tau}}(\underline{z})$ have GMLR in $\underline{z}$. Then all monotone selection procedures form an essentially complete class in $\mathcal{C}$.

3.4.6 Example. Let $X_{i1}, \ldots, X_{in_i}$ be independently distributed normal random variables with mean θ_i and variance $\sigma^2 = 1$, $i = 1, 2, \ldots, k$. Then the likelihood function of $\underline{\theta}$ is

$$g_{\underline{\theta}}(\underline{z}) = \prod_{i=1}^{k} g_{\theta_i}(\bar{x}_i),$$

where

$$g_{\theta_i}(\bar{x}_i) = \frac{\sqrt{n_i}}{\sqrt{2\pi}} e^{-\frac{n_i}{2}(x_i - \theta_i)^2}, \quad \bar{x}_i = \frac{1}{n_i} \sum_{\ell=1}^{n_i} x_{i\ell}.$$

Let $\tau_{ij} = \theta_i - \theta_j$, $1 \leq i, j \leq k$; and $Z_{ij} = \bar{X}_i - \bar{X}_j$, $1 \leq i, j \leq k$. It can be seen that $\tau_i = \theta_i - \theta_{[k]}$ if $\theta_i < \theta_{[k]}$ and $\tau_i = \theta_{[k]} - \theta_{[k-1]}$ if $\theta_i = \theta_{[k]}$ where $\theta_{[1]} \leq \cdots \leq \theta_{[k]}$. Thus the population with the largest mean is the best. If, instead, $\tau_{ij} = \theta_j - \theta_i$, then the population with the smallest mean would be the best. In this case, we have $h(t) = -t$ which is decreasing. We know that $(\bar{X}_1, \ldots, \bar{X}_k)$ is a sufficient statistic for $\underline{\theta}' = (\theta_1, \ldots, \theta_k)$. Let G consist of the transformations

$$g_c(\bar{x}_1, \ldots, \bar{x}_k) = (\bar{x}_1 + c, \ldots, \bar{x}_k + c).$$

Then $\underline{Z}_i' = (Z_{i1}, \ldots, Z_{ik})$ is a maximal invariant. The induced group of transformations on the parameter space is then

$$\bar{G} = \{\bar{g}: \bar{g}(\theta_1, \ldots, \theta_k) = (\theta_1 + c, \ldots, \theta_k + c)\}.$$

Then $\underline{\tau}_i' = (\tau_{i1}, \ldots, \tau_{ik})$ is a maximal invariant under $\bar{G}$. The distribution of $\underline{Z}_i$ depends only on $\underline{\tau}_i$.

For any i and

$$\underline{z}_i' = (z_{i1}, \ldots, z_{ik}), \quad \underline{\tau}_i' = (\tau_{i1}, \ldots, \tau_{ik}) \in R^{k-1},$$

the joint density of Z_{ij}, $j \neq i$, is given by

$$p_{\underline{\tau}_i}(\underline{z}_i) = (2\pi)^{-\frac{1}{2}(k-1)}|\Sigma_i|^{-\frac{1}{2}} \exp\{-\frac{1}{2}(\underline{z}_i-\underline{\tau}_i)'\Sigma_i^{-1}(\underline{z}_i-\underline{\tau}_i)\},$$

where Σ_i $((k-1)\times(k-1)) = (\sigma_{rs}^i)$ is the covariance matrix of Z_{ij}'s, $1 \leq j \leq k$, $j \neq i$, σ_{rs}^i is the covariance of Z_{ir} and Z_{is} $(r, s \neq i)$, and

$$\sigma_{rr}^i = \frac{1}{n_i} + \frac{1}{n_r}, \quad r = 1,\ldots,k, \ r \neq i,$$

$$\sigma_{rs}^i = \frac{1}{n_i}, \quad \text{for all } r \neq s, \ r, s \neq i,$$

$$\Sigma_i^{-1} = (a_{rs}^i),$$

where $\sum_{i=1}^{k} n_i = N$, and

$$a_{rr}^i = n_r - \frac{n_r^2}{N}, \quad r = 1,\ldots,k, \ r \neq i,$$

$$a_{rs}^i = -\frac{n_r n_s}{N}, \quad r,s = 1,\ldots,k, \ r \neq s, \ r,s \neq i.$$

Now for all i, j, $(i \neq j)$, and for all fixed ℓ, $(\ell = 1,\ldots,k, \ \ell \neq i, j)$,

$$p_{i1}(\underline{z}_i) = (2\pi)^{-\frac{1}{2}(k-1)}|\Sigma_i|^{-\frac{1}{2}} \exp\{-\frac{1}{2}(\underline{z}_i-\underline{\tau}_i)'\Sigma_i^{-1}(\underline{z}_i-\underline{\tau}_i)\},$$

and

$$p_{i0}(\underline{z}_i) = (2\pi)^{-\frac{1}{2}(k-1)}|\Sigma_i|^{-\frac{1}{2}} \exp\{-\frac{1}{2}(\underline{z}_i-\underline{\xi}_i)'\Sigma_i^{-1}(\underline{z}_i-\underline{\xi}_i)\},$$

where $\underline{\xi}_i' = (\xi_{i1},\ldots,\xi_{ik})$ and $\tau_{i\ell} = \xi_{i\ell}$, $1 \leq \ell \leq k$, $\ell \neq i, j$ and $\tau_{ij} \geq \xi_{ij}$, hence the ratio

$$\frac{p_{i1}(\underline{z}_i)}{p_{i0}(\underline{z}_i)} = \exp\{\frac{1}{2}(z_{ij}-\xi_{ij})[-\frac{n_j n_1}{N} z_{i1} - \ldots + (n_j - \frac{n_j^2}{N})z_{ij} - \ldots - \frac{n_j n_k}{N} z_{ik}]\}$$

which is nondecreasing in z_{ij}. Hence $p_{\underline{\tau}_i}(\underline{z}_i)$ has GMLR and the conditions of Theorem 3.4.5 are satisfied.

Now, we define a special monotone selection procedure as follows. For $i = 1,\ldots,k$, let

$$\delta_i^0(\underline{z}_i) = \begin{cases} 1 & \text{if } \underline{z}_i \geq \underline{d}_i, \\ 0 & \text{otherwise}, \end{cases}$$

where $\underline{d}_i = (d_{i1},\ldots,d_{ik})$, satisfies

$$\int \delta_i^0(\underline{z}_i) p_i(\underline{z}_i) d\nu(\underline{z}_i) = \gamma_i.$$

Note that if we formulate the selection problem as the case of hypotheses testing as follows:

$$H_0: \theta_1 = \ldots = \theta_k, \ H_i: \theta_i > \theta_j \ \forall j \neq i, \ i = 1,\ldots,k,$$

then γ_i is the error to choose H_i while H_0 is true, and γ_i is the so-called the error of i-th kind. Equivalently, in terms of $\underline{x} \in R^k$ this procedure is of the form

(3.4.6) $$\delta_i^0(\underline{x}) = \begin{cases} 1 & \text{if } \bar{x}_i \geq \max_{j \neq i}(\bar{x}_j + d_{ij}), \\ 0 & \text{otherwise.} \end{cases}$$

It should be pointed out that when all d_{ij}'s are negative, the monotone selection procedure $\delta^0 = (\delta_1^0, \ldots, \delta_k^0)$ given in (3.4.6) is the usual Gupta type procedure (cf. Gupta (1965)) to select a subset containing the best population associated with the largest θ as follows:

$$\delta_i^0(\underline{z}) = \begin{cases} 1 & \text{if } \bar{x}_i \geq \max_{1 \leq j \leq k}(\bar{x}_j - (-d_{ij})), \\ 0 & \text{otherwise.} \end{cases}$$

Gupta and Huang (1976) have studied the selection rule for the k normal means problem with a common known variance σ^2 based on samples of unequal sizes. In their solution the monotone rules are given by

$$d_{ij} = -d\sigma\sqrt{\frac{1}{n_i} + \frac{1}{n_j}}, \quad d > 0.$$

Hence this rule belongs to the essentially complete class and is optimal in this sense.

Gupta and Huang (1976) have computed the constant d for given P* to satisfy the following condition:

(3.4.7) $$\inf_{\theta \in \Omega} P_\theta(CS|\delta^0) = P^*.$$

Let $\bar{X}_{(i)}$ and $n_{(i)}$ denote the sample mean and the sample size associated with the population $\pi_{(i)}$ with mean $\theta_{[i]}$, $i = 1, 2, \ldots, k$ and $\sigma^2 = 1$. Of course, both $\bar{X}_{(i)}$ and $n_{(i)}$ are unknown. Then

$$P(CS|\delta^0) = P\{\bar{X}_{(k)} \geq \max_{1 \leq j \leq k-1}(\bar{X}_{(j)} - d\sqrt{\frac{1}{n_{(k)}} + \frac{1}{n_{(j)}}})\}$$

$$= P\{(\bar{X}_{(j)} - \bar{X}_{(k)})(\frac{1}{n_{(j)}} + \frac{1}{n_{(k)}})^{-\frac{1}{2}} \leq d, \quad j = 1, \ldots, k-1\}$$

$$= P\{(\bar{X}_{(j)} - \bar{X}_{(k)} - \mu_{[j]} + \mu_{[k]})(\frac{1}{n_{(j)}} + \frac{1}{n_{(k)}})^{-\frac{1}{2}}$$

$$\leq d + (\theta_{[k]} - \theta_{[j]})(\frac{1}{n_{(j)}} + \frac{1}{n_{(k)}})^{-\frac{1}{2}}, \quad j = 1, \ldots, k-1\}$$

(3.4.8) $$= P\{Z_{j,k} \leq d + (\theta_{[k]} - \theta_{[j]})(\frac{1}{n_{(j)}} + \frac{1}{n_{(k)}})^{-\frac{1}{2}}, \quad j = 1, \ldots, k-1\}.$$

For $\ell = 1, 2, \ldots, k$, define

(3.4.9) $\quad Z_{r,\ell} = (\bar{X}_{(r)} - \bar{X}_{(\ell)} - \theta_{[r]} + \theta_{[\ell]})(\frac{1}{n_{(r)}} + \frac{1}{n_{(\ell)}})^{-\frac{1}{2}}$, $r = 1, \ldots, k$; $r \neq \ell$,

and

(3.4.10) $\quad \rho_{r,s}^{(\ell)} = \rho(Z_{r,\ell}, Z_{s,\ell}) = [(1 + \frac{n_{(\ell)}}{n_{(r)}})(1 + \frac{n_{(\ell)}}{n_{(s)}})]^{-\frac{1}{2}}$, $r, s = 1, \ldots, k, r, s \neq \ell$; $r \neq s$.

Thus $Z_{r,\ell}$, $r \neq \ell$, are standard normal variables with correlation matrix $\{\rho_{r,s}^{(\ell)}\}$. We can write $P(CS|\delta^0)$ alternatively as

(3.4.11) $\quad P(CS|\delta^0) = \int_{-\infty}^{\infty} \prod_{j=1}^{k-1} \Phi\{\sqrt{\frac{n_{(j)}}{n_{(k)}}} y + \delta_{k,j} \sqrt{n_{(j)}} d + \sqrt{1 + \frac{n_{(j)}}{n_{(k)}}}\} d\Phi(y)$,

where $\delta_{ij} = \theta_i - \theta_j$, $\Phi(\cdot)$ denotes the cdf of a standard normal random variable.

For the evaluation of the infimum of $P(CS|\delta^0)$ over the parameter space

$$\Omega = \{\underline{\theta}: \underline{\theta} = (\theta_1, \ldots, \theta_k), -\infty < \theta_1, \ldots, \theta_k < \infty\}$$

and all possible associations between $(n_1, \ldots, n_k)$ and $(n_{(1)}, \ldots, n_{(k)})$, we need Slepian's theorem (see Section 1.3). For any ℓ, $1 \leq \ell \leq k$, let

(3.4.12) $\quad \kappa_{ij}^{(\ell)} = [(1 + \frac{n_{[\ell]}}{n_{[i]}})(1 + \frac{n_{[\ell]}}{n_{[j]}})]^{-\frac{1}{2}}$, $i, j = 1, \ldots, k$; $i, j \neq \ell$, $i \neq j$,

(3.4.13) $\quad \kappa_{ii}^{(\ell)} = 1$, $i = 1, \ldots, k$; $i \neq \ell$,

where $n_{[1]} \leq n_{[2]} \leq \ldots \leq n_{[k]}$ denote the ordered values of a given set $n_1, \ldots, n_k$ of positive integers. If $\underline{X}$ is a multivariate normal with means zero and correlation matrix $\{\kappa_{ij}^{(\ell)}\}$, for any fixed ℓ, $1 \leq \ell \leq k-1$ and $\underline{Y}$ is a multivariate normal with means zero and correlation matrix $\{\kappa_{ij}^{(k)}\}$, then we have

(3.4.14) $\quad P(\underline{X} \leq \underline{a}) = \Phi_{k-1}(a_1, \ldots, a_{k-1}; \{\kappa_{ij}^{(\ell)}\}) \geq \Phi_{k-1}(a_1, \ldots, a_{k-1}; \{\kappa_{ij}^{(k)}\}) = P(\underline{Y} \leq \underline{a})$.

The inequality (3.4.14) follows from Slepian theorem, if we show that $n_{ij} \geq \xi_{ij}$, $1 \leq i \leq k-1$, $1 \leq j \leq k-1$, where for any fixed ℓ, $(1 \leq \ell \leq k-1)$,

$$n_{ij} = \begin{cases} \kappa_{ij}^{(\ell)}, & 1 \leq i \leq \ell-1, 1 \leq j \leq \ell-1, \\ \kappa_{i,j+1}^{(\ell)}, & 1 \leq i \leq \ell-1, \ell \leq j \leq k-1, \\ \kappa_{i+1,j}^{(\ell)}, & \ell \leq i \leq k-1, 1 \leq j \leq \ell-1, \\ \kappa_{i+1,j+1}^{(\ell)}, & \ell \leq i \leq k-1, \ell \leq j \leq k-1, \end{cases}$$

and $\quad \xi_{ij} = \kappa_{ij}^{(k)}$, $1 \leq i \leq k-1$, $1 \leq j \leq k-1$.

Since for $1 \leq i \leq k-1$, $n_{[i]} \leq n_{[i+1]} \leq n_{[k]}$, it follows from (3.4.12) that, for any fixed ℓ, $(1 \leq \ell \leq k-1)$,

$$\kappa_{i+1,j}^{(\ell)} \geq \kappa_{ij}^{(\ell)} \geq \kappa_{ij}^{(k)} \text{ for all } i, j; i, j \neq \ell; i, j \neq k.$$

Also, since the correlation matrices are symmetric, the above inequalities imply that $n_{ij} \geq \xi_{ij}$ for all i,j ($1 \leq i, j \leq k-1$). This completes the proof.

For any given association between $(n_1,\ldots,n_k)$ and $(n_{(1)},\ldots,n_{(k)})$, we can see from (3.4.11) that the infimum of $P(CS|\delta^0)$ is attained when $\theta_{[1]} = \ldots = \theta_{[k]}$. Thus the infimum we seek in (3.4.7) is given by

$$(3.4.15) \quad \min_{1 \leq i \leq k} \int_{-\infty}^{\infty} \prod_{\substack{j=1 \\ j \neq i}}^{k} \Phi\left\{\sqrt{\frac{n_{(j)}}{n_{(i)}}}\, y + d\sqrt{1 + \frac{n_{(j)}}{n_{(i)}}}\right\} d\Phi(y) = P^*.$$

Using the alternative form in (3.4.8), this minimum in (3.4.15) is equal to

$$\min_{1 \leq \ell \leq k} P\{Z_{r,\ell} \leq d, \; r = 1,\ldots,k; \; r \neq \ell\}$$

$$= \min_{1 \leq \ell \leq k} \Phi_{k-1}(d,\ldots,d; \{\rho_{r,s}^{(\ell)}\})$$

$$= \min_{1 \leq \ell \leq k} \Phi_{k-1}(d,\ldots,d; \{\kappa_{r,s}^{(\ell)}\})$$

$$= \Phi_{k-1}(d,\ldots,d; \{\kappa_{r,s}^{(k)}\})$$

$$= P\{V_j \leq d, \; i = 1,\ldots,k-1\},$$

where $V_1,\ldots,V_{k-1}$ are standard normal random variables with correlation $\kappa_{r,s}^{(k)} = \alpha_r \alpha_s$, where

$$\alpha_i = \left(1 + \frac{n_{[k]}}{n_{[i]}}\right)^{-\frac{1}{2}}, \; i = 1,\ldots,k-1.$$

It is well known (see Section 1.3) that $V_1,\ldots,V_{k-1}$ can be generated from k independent standard normal variates $V_1',\ldots,V_{k-1}'$, V by the transformation

$$V_j = (1-\alpha_j^2)^{\frac{1}{2}} V_j' + \alpha_j V$$

and it follows that

$$(3.4.16) \quad \inf P(CS|\delta^0) = \int_{-\infty}^{\infty} \prod_{j=1}^{k-1} \Phi\left[\frac{d - \alpha_j u}{(1-\alpha_j^2)^{\frac{1}{2}}}\right] d\Phi(u).$$

For convenience in using the existing tables, we compute a lower bound for (3.4.16) as follows. Since

$$\int_{-\infty}^{\infty} \prod_{j=1}^{k-1} \Phi\left[\frac{d - \alpha_j u}{(1-\alpha_j^2)^{\frac{1}{2}}}\right] d\Phi(u)$$

$$\geq \int_{-\infty}^{\infty} \Phi^{k-1}\left(\frac{d - \rho^{\frac{1}{2}} u}{\sqrt{1-\rho}}\right) d\Phi(u).$$

Equating the above integral to P^*, values of d are available for the equicorrelated random variables from the tables in Gupta, Nagel and Panchapakesan (1973). These constant values will be greater than the exact values satisfying the equation (3.4.16), and these values will lead to exact values in the case of $n_{[1]} = \ldots = n_{[k]}$. Some of the exact d-values can be obtained from Table 1 of Gupta and Huang (1976).

It should be pointed out that the Seal-type procedure (Seal, (1955)) for the above normal means example is also a monotone selection procedure. The procedure has been proposed by Seal (1955) as follows.

R_s: Select π_i if and only if $\bar{X}_i \geq \frac{1}{k-1} \sum_{j \neq i} \bar{X}_j - C_i$.

We can rewrite it in terms of Z_{ij}'s as follows.

R_s: Select π_i if and only if $\sum_{j \neq i} Z_{ij} \geq -c_i(k-1)$.

By Definition 3.4.4, R_s is a monotone procedure.

Note that the two rules of the above example have been compared for their performance in terms of the expected size of the selected subset. Bayes optimality and other optimality criteria have been considered by Deely and Gupta (1968), Gupta and Hsu (1978) and Gupta and Miescke (1978). The Gupta type rule performs better than Seal's average-type rule R_s, in general, if the means are not very close. Also the Gupta-type rules provide good approximations for the Bayesian rules under exchangeable normal priors. This was demonstrated theoretically as well as numerically via simulation studies in the papers mentioned above.

Berger and Gupta (1980) have studied a different criterion of optimality for the selection problems as follows:

We will consider only those decision rules φ which satsify the P*-condition,

$$\inf_{\theta \in \Theta} P_\theta(CS|\varphi) \geq P^*.$$

Let $\pi_{(1)}, \ldots, \pi_{(k-1)}$ be the non-best populations with associated parameters $\theta_{[1]}, \ldots, \theta_{[k-1]}$. We use the risk function to reflect the fact that the non-best populations are included with "low" probability. This function is as follows:

$$M(\underline{\theta}, \varphi) = \max_{1 \leq i \leq k-1} P_\theta(\text{select } \pi_{(i)}|\varphi).$$

For the normal means problem, the following selection rule has been proposed by Gupta (1965) for the case $n_1 = \ldots = n_k = n$.

R_1: Select π_i if and only if $\bar{x}_i \geq \max_{1 \leq j \leq k} \bar{x}_j - d$.

Berger and Gupta (1980) have shown that R_1 is minimax with respect to M in the class of non-randomized, just, and translation invariant rules which satisfy the P*-condition. Furthermore, R_1 is the unique minimax rule in this class so R_1 is admissible in this class.

Furthermore, Berger and Gupta (1980) have proved the selection rule δ^0 (see (3.4.6)) is minimax with respect to M in the class of non-randomized, 'just', and translation invariant rules which satisfy the P*-condition. Note that the rule δ^0

applies to the unequal sample sizes (unequal variances) case.

A different type of essentially complete class of decision rules for the selection problems will be studied below.

Let X be a real random observation whose density is $f_\theta(x)$ and $F_\theta(x) = \int_{-\infty}^{x} f_\theta(u)dv(u)$, for $\theta \in \Theta$, where Θ is a subset of the real line and ν is a σ-finite, nonatomic measure. A set function μ defined on a Borel field $\mathcal{B}$ is called nonatomic if it has the following property: If for some $S \in \mathcal{B}$, $\mu(S) \neq 0$, then there exists an $S' \subset S$ such that $S' \in \mathcal{B}$ and such that $\mu(S') \neq \mu(S)$ and $\mu(S') \neq 0$. A cumulative distribution function is called nonatomic if its associated set function is nonatomic.

We assume $f_\theta(x)$ has strict monotone likelihood ratio. We are concerned with inference on θ (or with a monotone function of θ). The action space is a subset of the real line and the loss function is denoted by $L(\theta,a)$ for $a \in \mathcal{Q}$, the action space. For convenience we take $\mathcal{Q} = \Theta$, although the more general case can be treated similarly. Brown, Cohen and Strawderman (1976) give the following definition and condition: The loss function $L(\theta,a)$ is defined to be bowl shaped, if for each fixed θ, $L(\theta,\cdot)$ is nonincreasing for $a < \theta$, $L(\theta,\theta) = 0$, and $L(\theta,\cdot)$ is nondecreasing for $a > \theta$.

3.4.7 <u>Condition</u>. For every action a, either

(i) there exists a θ, say θ_a, with $\theta_a \leq a$, such that for every action $a' > a$,
$L(\theta_a, a') - L(\theta_a, a) > 0$ or

(ii) there exists a θ, say θ_a, with $\theta_a \geq a$, such that for every action $a' < a$,
$L(\theta_a, a') - L(\theta_a, a) > 0$.

3.4.8 <u>Definition</u>. A procedure δ, is called monotone if for any $x < y$, and any $a \in \mathcal{Q}$, the condition $\delta(y; (-\infty,a]) > 0$ implies that $\delta(x; (a,\infty)) = 0$.

Note that a more general definition of monotone procedures is given by Karlin and Rubin (1956).

We will assume that $L(\theta,a)$ is lower semicontinuous, bowl shaped, and satisfies Condition 3.4.7. Let $R(\theta,\delta)$ denote the risk function for a procedure δ. Also let F_θ^{-1} denote the left continuous inverse function of F_θ. Brown, Cohen and Strawderman (1976) have proved the following theorem.

3.4.9 <u>Theorem</u>. Let δ be any given nomonotone procedure. Let b be any real number and let $K(A,\delta) = \int \delta(x; A) f_\theta(x) dv(x)$. Define δ' as follows:

$$\delta'(x; (-\infty,b]) = 1 \quad \text{if} \quad x \leq F_b^{-1}(K_b(-\infty,b],\delta)),$$

$$= 0 \quad \text{if} \quad x > F_b^{-1}(K_b((-\infty,b),\delta)).$$

Then $R(\theta,\delta') \leq R(\theta,\delta)$ for all $\theta \in \Theta$ with strict inequality for some $\theta \in \Theta$.

Oosterhoff (1969) defines a strict monotone likelihood ratio density for a random vector $\underline{X}$ of order k×1 as follows: Let $\underline{\theta}$ be a k×1 vector of parameters. A partial ordering of points in $\mathbb{R}^k$ is defined by $\underline{x}_1 \leq \underline{x}_2$, meaning $x_{1i} \leq x_{2i}$, for $i = 1,2,\ldots,k$, and the inequality is strict for at least one component, and $\underline{x}_1 < \underline{x}_2$ means $x_{1i} < x_{2i}$ for all i, $1 \leq i \leq k$.

3.4.10 Definition. The density $f(\underline{x}; \underline{\theta})$ has a strict monotone likelihood ratio if for $\underline{\theta}_1 < \underline{\theta}_2$, $[f(\underline{x}; \underline{\theta}_2)/f(\underline{x}; \underline{\theta}_1)]$ is strictly increasing in $\underline{x}$.

Oosterhoff (1969) considers the hypothesis testing problem H_0: $\underline{\theta} = \underline{\theta}_0$ vs H_1: $\underline{\theta}_0 < \underline{\theta}$. He proves that if $f(\underline{x}; \underline{\theta})$ has strict monotone likelihood ratio, then for the hypothesis testing problem H_0 vs H_1, the class of monotone procedures is essentially complete. Brown, Cohen and Strawderman (1976) prove that the class of monotone procedures is complete. Earlier Karlin and Rubin (1956) proved that if $f_\theta(x)$ has monotone likelihood ratio, then the class of monotone procedures is essentially complete.

References

[1] Berger, R. L. (1977). Minimax subset selection for loss measured by subset size. Mimeo. Ser. #489, Dept. of Statist., Purdue Univ.

[2] Berger, R. L. and Gupta, S. S. (1980). Minimax subset selection rules with applications to unequal variance (unequal sample size) problems. Scand. J. Statist. 7, 21-26.

[3] Bjørnstad, J. (1980). A class of Schur-procedures and minimax theory for subset selection. To appear in Ann. Statist.

[4] Brown, L. D., Cohen, A. and Strawderman, W. E. (1976). A complete class of theorem for strict monotone likelihood ratio with applications. Ann. Statist. 4, 712-722.

[5] Deely, J. J. and Gupta, S. S. (1968). On the properties of subset selection properties of subset selection procedures. Sankhyā Ser. A, 30, 37-50.

[6] Ferguson, T. S. (1967). Mathematical Statistics: A Decision Theoretic Approach. Academic Press, New York.

[7] Gupta, S. S. (1965). On multiple decision (selecting and ranking) rules. Technometrics 7, 225-245.

[8] Gupta, S. S. and Hsu, J. C. (1978). On the performance of some subset selection procedures. Comm. Statist. B7, 561-591.

[9] Gupta, S. S. and Huang, D. Y. (1976). Subset selection procedure for the means and variances of normal populations: unequal sample sizes case. Sankhyā Ser. B, 38, 112-128.

[10] Gupta, S. S. and Huang, D. Y. (1977). On some Γ-minimax selection and multiple comparison procedures. Statistical Decision Theory and Related Topics II (Ed. Gupta, S. S. and Moore, D. S.), 139-155.

[11] Gupta, S. S. and Huang, D. Y. (1980a). A note on optimal subset selection procedures. To appear in Ann. Statist.

[12] Gupta, S. S. and Huang, D. Y. (1980b). An essentially complete class of multiple decision procedures. To appear in Journal of Statistical Planning and Inference.

[13] Gupta, S. S. and Miescke, K. J. (1978). Optimality of subset selection procedure for ranking means of three normal populations. Mimeo. Ser. #78-19, Dept. of Statist., Purdue Univ., Lafayette, IN 47907.

[14] Gupta, S. S., Nagel K. and Panchapakesan, S. (1973). On the order statistics from equally correlated normal variables. Biometrika 60, 403-413.

[15] Karlin, S. and Rubin, H. (1956). The theory of decision procedures for distributions with monotone likelihood ratio. Ann. Math. Statist. 27, 272-299.

[16] Lehmann, E. L. (1961). Some model I problems of selection. Ann. Math. Statist. 32, 990-1012.

[17] Miescke, K. J. (1979). Γ-minimax selection procedures in simultaneous testing problems. Mimeo. Ser. #79-1, Dept. of Statist., Purdue Univ., IN 47907.

[18] Nagel, K. (1970). On subset selection rules with certain optimality properties. Mimeo. Ser. #222, Dept. of Statist., Purdue Univ., Lafayette, IN 47907.

[19] Oosterhoff, J. (1969). Combination of One-Sided Statistical Tests. Mathematical Centre Tracts 28, Amsterdam.

[20] Randles, R. H. and Hollander, M. (1971). Γ-minimax selection procedures in treatments versus control problems. Ann. Math. Statist. 42, 330-341.

[21] Seal, K. C. (1955). On a class of decision procedures for ranking means of normal populations. Ann. Math. Statist. 36, 387-397.

CHAPTER 4
INVARIANT DECISION PROCEDURES

4.1 Introduction

In Section 4.2, we define and study invariant procedures. We describe optimal rules in the class of invariant procedures. Various optimality criteria of the goodness of a rule are mentioned. Various selection procedures based on sample values and their ranks are also studied. Different types of loss functions for subset selection approach for the derivation of selection rules are described and studied. Optimality of classical selection rules in terms of their closeness to the Bayes rules is considered.

4.2 Selecting the Best Population

The problem of selecting the best one of several populations is a multiple decision problem with k possible decisions. More generally, let us consider a multiple decision problem with possible decisions $d_1,\ldots,d_k$. We define a (randomized) decision procedure as a vector $\varphi = (\varphi_1,\ldots,\varphi_k)$ where $\varphi_i(\underline{x})$ denotes the probability of making decision d_i when the value of the random observable $\underline{X}$ is $\underline{x}$, and we assume that $\sum_1^k \varphi_i(\underline{x}) = 1$ for all $\underline{x}$. Suppose that the distribution P_θ of $\underline{X}$ depends on the parameter $\underline{\theta}$ and that the loss resulting from the decision d_i when $\underline{\theta}$ is the true parameter value is $L(\underline{\theta},d_i)$, and the risk is $R(\underline{\theta},d_i) = EL(\underline{\theta},d_i(\underline{X}))$.

We restrict attention to the symmetric case where typically the observations consist of samples of equal size from different populations. We assume that the problem is invariant under a finite transformation group $G = \{g_1,\ldots,g_N\}$: if the distribution of $\underline{X}$ is $P_{\underline{\theta}}$, the random vector $g_i \underline{X}$ has distribution $P_{\bar{g}_i\underline{\theta}}$ where g_i and $\bar{g}_i$ are one to one mappings, respectively, of the sample space and of the parameter space onto themselves; furthermore, there exist transformations $g_1^*,\ldots,g_N^*$ of the decision space (i.e. permutations of $d_1,\ldots,d_k$) such that for any i, j and $\underline{\theta}$

(4.2.1) $\qquad L(\bar{g}_i\underline{\theta}, g_i^* d_j) = L(\underline{\theta},d_j)$.

4.2.1 Definition. A procedure is said to be invariant if
(4.2.2) $\qquad g^*\varphi(\underline{x}) = \varphi(g\underline{x})$ for all $\underline{x}$ and g.

Note that the procedure taking on the value $\varphi(g\underline{x})$ at the point $\underline{x}$ will be denoted by φg, and (4.2.2) can then be written as $g^*\varphi g^{-1} = \varphi$.

4.2.2 Definition. Suppose an ordering "$\leq$" is introduced among the procedures. A procedure φ' is "at least as good as" φ if $\varphi \leq \varphi'$.

Lehmann (1966) has discussed some results for optimal procedures using various different criteria as follows:

4.2.3 **Lemma**. Suppose that the ordering criterion satisfies $\sup R(\underline{\theta},\varphi') \leq \sup R(\underline{\theta},\varphi)$. Then we define $\varphi \leq \varphi'$. Then,

(a) given any procedure φ, there exists an invariant procedure φ' such that $\varphi \leq \varphi'$ and

(b) if there exists a procedure $\varphi^{(0)}$ that uniformly minimizes the risk among all invariant procedures, $\varphi^{(0)}$ is optimum with respect to the ordering $\leq$, i.e.,

$$\varphi \leq \varphi^{(0)} \text{ for all } \varphi.$$

Proof.
(a) Let $\varphi' = \frac{1}{N} \sum_1^N g_i^* \varphi g_i^{-1}$. Since

$$g^* \varphi'(\underline{x}) = \frac{1}{N} \sum_1^N g^* g_i^* \varphi g_i^{-1}(\underline{x})$$

$$= \frac{1}{N} \sum_1^N (gg_i)^* \varphi (gg_i)^{-1}(g\underline{x})$$

$$= \frac{1}{N} \sum_1^N g_i^* \varphi g_i^{-1}(g\underline{x}) = \varphi'(g\underline{x}),$$

hence φ' is invariant and

$$\sup_{\underline{\theta}} R(\underline{\theta},\varphi') \leq \frac{1}{N} \sum_1^N \sup_{\underline{\theta}} R(\underline{\theta}, g_i^* \varphi g_i^{-1})$$

$$= \frac{1}{N} \sum_1^N \sup_{\underline{\theta}} R(\bar{g}_i\underline{\theta}, g_i^* \varphi g_i^{-1})$$

$$= \frac{1}{N} \sum_1^N \sup_{\underline{\theta}} R(\underline{\theta},\varphi) = \sup_{\underline{\theta}} R(\underline{\theta},\varphi).$$

(b) This is an immediate consequence of (a).

The following examples of some ordering "$\leq$", in general, are discussed by Lehmann (1966).

4.2.4 **Example**. Minimax: $\varphi \leq \varphi'$ if $\sup_{\underline{\theta}} R(\underline{\theta},\varphi') \leq \sup_{\underline{\theta}} R(\underline{\theta},\varphi)$.

4.2.5 **Example**. Minimax regret: For a multiple decision problem this ordering is given by $\varphi \leq \varphi'$ if

$$\sup_{\underline{\theta}} [R(\underline{\theta},\varphi') - \min_i L_i(\underline{\theta})] \leq \sup_{\underline{\theta}} [R(\underline{\theta},\varphi) - \min_i L_i((\underline{\theta}))].$$

4.2.6 **Example**. Average risk: $\varphi \leq \varphi'$ if for some specified $\underline{\theta}_0$

$$\frac{1}{N} \sum_1^N R(\bar{g}_i\underline{\theta}_0,\varphi') \leq \frac{1}{N} \sum_1^N R(\bar{g}_i\underline{\theta}_0,\varphi).$$

Lehmann (1966) has provided an important principle to construct optimal procedures. The result is as follows:

4.2.7 Theorem. A necessary and sufficient condition for an invariant procedure $\varphi^{(0)}$ to minimize the risk $R(\underline{\theta},\varphi)$ among all invariant procedures is that it minimizes the average risk

$$(4.2.3) \qquad \gamma(\underline{\theta},\varphi) = \frac{1}{N} \sum_{1}^{N} R(\bar{g}_i \underline{\theta},\varphi)$$

among all procedures.

Proof. (i) Let $\varphi^{(0)}$ be an invariant procedure which minimizes $\gamma(\underline{\theta},\varphi)$ among all procedures. If φ' is any other invariant procedure, it follows from the fact that the risk function of any invariant procedure is constant over each orbit $\{\bar{g}_i \underline{\theta}: i = 1,\ldots,N\}$ (see Ferguson (1967), p. 149) that

$$R(\underline{\theta},\varphi') = \gamma(\underline{\theta},\varphi') \geq \gamma(\underline{\theta},\varphi^{(0)}) = R(\underline{\theta},\varphi^{(0)}).$$

Hence $\varphi^{(0)}$ minimizes $R(\underline{\theta},\varphi)$ among all invariant procedures.

(ii) Suppose, conversely, that $\varphi^{(0)}$ minimizes $R(\underline{\theta},\varphi)$ among all invariant procedures and let φ' be any procedure. Then there exists an invariant procedure φ'' such that $\gamma(\underline{\theta},\varphi'') \leq \gamma(\underline{\theta},\varphi')$. This follows from Lemma 4.2.3. We now have

$$\gamma(\underline{\theta},\varphi') \geq \gamma(\underline{\theta},\varphi'') = R(\underline{\theta},\varphi'') \geq R(\underline{\theta},\varphi^{(0)}) = \gamma(\underline{\theta},\varphi^{(0)}).$$

This completes the proof.

Based on the criteria described above, Bahadur and Goodman (1952) derive an optimal procedure among invariant rules. Let the random observable $\underline{X}$ have a density of the form

$$(4.2.4) \qquad h_{\underline{\theta}}(\underline{t}) = c(\underline{\theta}) f_{\theta_1}(t_1) \ldots f_{\theta_k}(t_k)$$

with respect to some σ-finite measure μ on the sample space where $\underline{\theta} = (\theta_1,\ldots,\theta_k)$, $\underline{t} = (t_1,\ldots,t_k)$ and $t_i = t_i(\underline{x})$, $i = 1,\ldots,k$, are real-valued statistics. Suppose that f_{θ_i} has monotone (non-decreasing) likelihood ratio (MLR) in t_i for each $i = 1,\ldots,k$. Consider the problem of selecting the largest among $\theta_1,\ldots,\theta_k$ and let d_i denote the selection of θ_i.

We assume that $\underline{x} = (x_1,\ldots,x_n)$ and that there exist permutations of the x's which leave μ invariant. It is enough to reduce the data to the sufficient statistics $\underline{t} = (t_1,\ldots,t_k)$ whose joint density is given by (4.2.4) with respect to a σ-finite measure ν in t-space which we assume to be invariant under the group G of all permutations of $(t_1,\ldots,t_k)$.

Lehmann (1966) gives an alternative proof of the theorem of Bahadur and Goodman (1952) to obtain an optimal procedure as follows.

4.2.8 Theorem. Let the distribution of the sufficient statistics $T = (T_1,\ldots,T_k)$ have density (4.2.4) with respect to a σ-finite measure ν which is invariant under G. For any permutation g of $(t_1,\ldots,t_k)$ define g^- and g^+ as the same permutation of $(\theta_1,\ldots,\theta_k)$ and $(d_1,\ldots,d_k)$, respectively, and suppose that the loss function L satisfies (4.2.1) and

(4.2.5) $$\theta_i < \theta_j \Rightarrow L_i(\underline{\theta}) \geq L_j(\underline{\theta}).$$

Let $\varphi^{(0)}$ be the procedure which takes decision d_i when t_i is the unique largest among $(t_1,\ldots,t_k)$, and which takes decisions $d_{i_1},\ldots,d_{i_r}$ each with probability $\frac{1}{r}$ if $(t_{i_1},\ldots,t_{i_r})$ is the set of t-values equal to max t_j (i.e., which breaks ties at random). Then $\varphi^{(0)}$ uniformly minimizes the risk among all procedures based on t which are invariant under G.

Eaton (1967) has generalized the above results to a fairly general treatment of a class of ranking (or selection) problems. In his paper Eaton (1967) restricts attention to the symmetric case so that certain covariance arguments are applicable.

Eaton (1967) considers a random observable $\underline{Z} = (\underline{X},\underline{Y})$ with values in a measurable space $(\mathcal{X} \times \mathcal{Y}, \mathcal{B}(\mathcal{X}) \times \mathcal{B}(\mathcal{Y}))$ so that $\underline{X}$ has values in $\mathcal{X}$ and $\underline{Y}$ has values in $\mathcal{Y}$. The space $\mathcal{X}$ is assumed to be a symmetric Borel subset of $\mathbb{R}^k$ and $\mathcal{B}(\mathcal{X})$ is the Borel field inherited from $\mathbb{R}^k$ while $\mathcal{Y}$ is arbitrary. It is assumed that $\underline{Z}$ has a density

(4.2.6) $$p_{\underline{\alpha}}(\underline{x},\underline{y}; \underline{\theta})$$

with respect to a σ-finite measure μ on $\mathcal{B}(\mathcal{X}) \times \mathcal{B}(\mathcal{Y})$, where $(\underline{\theta},\underline{\alpha})$ is a parameter in a set $\Theta \times A$ ($\underline{\theta} \in \Theta$, $\underline{\alpha} \in A$). The set Θ is assumed to be a symmetric Borel subset of $\mathbb{R}^k$. For the ranking problem, we are trying to rank θ_i's while $\underline{\alpha}$ is a nuisance parameter vector. In most problems, the observation $\underline{Z}$ represents a sufficient statistic for the parameter $(\underline{\theta},\underline{\alpha})$ based on samples of size n from each of k populations.

Given the above structure, the ranking problem may be described as follows: on the basis of $\underline{Z}$, partition the set of coordinate values of the parameter $\underline{\theta} = (\theta_1,\ldots,\theta_k)$ into s disjoint subsets, say $\lambda_1,\ldots,\lambda_s$, such that λ_1 contains the k_1 largest θ_i, λ_2 contains the k_2 next largest $\theta_i,\ldots,$ and λ_s contains the k_s smallest θ_i where $1 \leq k_i < k$ and $\sum_{i=1}^{s} k_i = k$. An equivalent formulation of the above problem is: partition the set $\{1,2,\ldots,k\}$ into s disjoint subsets, say $\gamma_1,\ldots,\gamma_s$, where γ_i has k_i elements, $1 \leq k_i < k$, $\sum_{1}^{k} k_i = k$ and then make the obvious association between γ_i and λ_i. Thus the action space for the ranking problem can be taken to be the set $\pi_s = \{\gamma\}$ of all partitions $\gamma = (\gamma_1,\ldots,\gamma_s)$ of $\{1,2,\ldots,k\}$ where γ_i has k_i elements and the k_i are fixed, $1 \leq k_i < k$, $\sum_{1}^{s} k_i = k$.

As before in Lehmann's (1966) treatment of the selection problem, invariance plays a central role in the treatment here. Let π denote a permutation of the set $\{1,2,\ldots,k\}$ and let G be the group of such permutations. The element of G which interchanges i and j, leaving all other members of $\{1,2,\ldots,k\}$ fixed, is denoted by (i,j). For $(\underline{x},\underline{y}) \in \mathcal{X} \times \mathcal{Y}$ and $\pi \in G$, define $\pi(\underline{x},\underline{y})$ by $\pi(\underline{x},\underline{y}) \equiv (\pi\underline{x},\underline{y})$ where $\pi\underline{x}$ is defined by $(\pi\underline{x})_i \equiv x_{\pi^{-1}i}$. With this definition, it is easy to check that

$(\pi_1\pi_2)(\underline{x},\underline{y}) = \pi_1(\pi_2(\underline{x},\underline{y}))$ so that the group G operates on the left of the space $\mathcal{X} \times \mathcal{Y}$. Similarly, for $(\underline{\theta},\underline{\alpha}) \in \Theta \times A$ and $\pi \in G$, $\pi(\underline{\theta},\underline{\alpha})$ is defined by $\pi(\underline{\theta},\underline{\alpha}) \equiv (\pi\underline{\theta},\underline{\alpha})$ where $(\pi\underline{\theta})_i \equiv \theta_{\pi^{-1}i}$. Also, for $\gamma = (\gamma_1,\ldots,\gamma_s) \in \pi_s$ and $\pi \in G$, define $\pi\gamma$ by $\pi\gamma \equiv (\pi\gamma_1,\ldots,\pi\gamma_s)$ where $\pi\gamma_i$ is the image of γ_i under π. For the density p and the measure μ, the following invariance is assumed:

(4.2.7) $\qquad p_{\underline{\alpha}}(\underline{x},\underline{y}; \underline{\theta}) = p_{\underline{\alpha}}(\pi\underline{x},\underline{y}; \pi\underline{\theta})$,

(4.2.8) $\qquad d\mu(\underline{x},\underline{y}) = d\mu(\pi\underline{x},\underline{y})$.

A decision function φ for this ranking problem is a measurable vector function on $\mathcal{X} \times \mathcal{Y}$ such that $\varphi = \{\varphi_\gamma : \gamma \in \pi_s\}$ where $0 \leq \varphi \leq 1$ and $\sum_{\gamma \in \pi_s} \varphi_\gamma = 1$. Let $\mathcal{D}$ be the class of such decision functions. For $\varphi \in \mathcal{D}$ and $\pi \in G$, define $\pi\varphi$ by $(\pi\varphi)_\gamma = \varphi_{\pi^{-1}\gamma}$. A decision φ is invariant if $\pi\varphi(\underline{x},\underline{y}) = \varphi(\pi\underline{x},\underline{y})$, that is, if $\varphi_{\pi^{-1}\gamma}(\underline{x},\underline{y}) = \varphi_\gamma(\pi\underline{x},\underline{y})$. Let $\mathcal{D}_I$ be the set of invariant decision functions. To introduce the loss structure of the problem, let $L_\gamma(\underline{\theta},\underline{\alpha})$ be the loss in taking action $\gamma \in \pi_s$ at the parameter point $(\underline{\theta},\underline{\alpha})$. Before stating the assumptions on the loss functions, we need the following definition (see Eaton (1967)).

4.2.9 **Definition.** If $\gamma = (\gamma_1,\ldots,\gamma_s)$ and $\gamma' = (\gamma_1',\ldots,\gamma_s')$ are elements of π_s, then γ differs adjacently from γ' at $[i,j]$ if there exists an integer β ($1 \leq \beta < s$) such that: (i) $i \in \gamma_\beta$, $i \in \gamma_{\beta+1}'$, (ii) $j \in \gamma_\beta'$, $j \in \gamma_{\beta+1}$, and (iii) $(i,j)\gamma' = \gamma$.

The loss function $L_\gamma(\underline{\theta},\underline{\alpha})$ are assumed to satisfy the following:

(4.2.9) If γ differs adjacently from γ' at $[i,j]$, then $L_\gamma(\underline{\theta},\underline{\alpha}) \leq L_{\gamma'}(\underline{\theta},\underline{\alpha})$ when $\theta_i \geq \theta_j$,

(4.2.10) $\qquad 0 \leq L_\gamma(\underline{\theta},\underline{\alpha}) = L_{\pi\gamma}(\pi\underline{\theta},\underline{\alpha})$.

The risk function of $\varphi \in \mathcal{D}$ is defined by

(4.2.11) $\quad \rho(\varphi,\underline{\theta},\underline{\alpha}) = \int \Sigma_\gamma \varphi_\gamma(\underline{x},\underline{y}) L_\gamma(\underline{\theta},\underline{\alpha}) p_{\underline{\alpha}}(\underline{x},\underline{y}; \underline{\theta}) d\mu(\underline{x},\underline{y})$.

Assuming that $p_{\underline{\alpha}}(\underline{x},\underline{y}; \underline{\theta})$ has property M for all $\underline{y} \in \mathcal{Y}$, Eaton (1967) has proved the result that for all $(\underline{\theta},\underline{\alpha}) \in \Theta \times A$,

(4.2.12) $\qquad \rho(\varphi^*,\underline{\theta},\underline{\alpha}) \leq (\varphi,\underline{\theta},\underline{\alpha})$ for all $\varphi \in \mathcal{D}_I$,

where the decision rule φ^* is defined by

(4.2.13) $\qquad \varphi_\gamma^*(\underline{x},\underline{y}) = \begin{cases} \dfrac{1}{n(\underline{x})} & \text{if } \gamma \in H(\underline{x}), \\ 0 & \text{if } \gamma \notin H(\underline{x}), \end{cases}$

and $H(\underline{x})$ is defined as follows: Let

(4.2.14) $\qquad B_\gamma = \{\underline{x} | \underline{x} \in \mathcal{X},\ x_{i_1} \geq \ldots \geq x_{i_s}$ for all $i_j \in \gamma_j,\ j = 1,\ldots,s\}$, then

$\qquad H(\underline{x}) = \{\gamma | \gamma \in \pi_s,\ \underline{x} \in B_\gamma\}$ for each $\underline{x} \in \mathcal{X}$.

Now let $n(\underline{x})$ be the number of elements in the set $H(\underline{x})$ so that $n(\underline{x}) \geq 1$.

Note that $\varphi^* = \{\varphi^*_\gamma | \gamma \in \pi_s\}$ is not a function of $\underline{y}$ and it is easy to see that $\varphi^* \in \mathcal{D}_I$.

Further, using Lemma 4.2.3, Eaton (1967) shows that φ^* is minimax and admissible. He also discusses some examples for certain specific applications of the results previously established as follows:

4.2.10 Example. Ranking Main Effects in Analysis of Variance.

For a fixed i consider independent observations

(4.2.15) $$y_{ij} = \beta_i + \xi_{ij} + \epsilon_{ij}$$

where $i = 1,\ldots,K$, $j = 1,\ldots,J$. The vector of parameters $\underline{\beta}' = (\beta_1,\ldots,\beta_K)$ is assumed to lie in a symmetric subset of $\mathbb{R}^k$. Let $\delta_i \equiv \frac{1}{J} \sum_{j=1}^{J} \xi_{ij}$ and assume that $\delta_1 = \ldots = \delta_K = \delta$. The random variables ϵ_{ij} have a joint normal distribution with mean $\underline{0}$ and covariance matrix $\Sigma = \sigma^2[(1-\rho)I+\rho\underline{e}\underline{e}']$ where I is the KJ × KJ identity matrix, $\underline{e}$ is the vector of 1's, σ^2 is the common variance of the ϵ_{ij} and ρ, $(-1/(KJ-1) < \rho < 1)$ is the correlation between any two different ϵ_{ij}.

Consider the problem of ranking the β_i with action space $\pi_s = \{\gamma\}$ and loss functions $L_\gamma(\underline{\beta})$ which satisfy (4.2.9) and (4.2.10) and $L_\gamma(\beta_1+c,\ldots,\beta_K+c) = L_\gamma(\beta_1,\ldots,\beta_K)$ for any real number c. To transform the problem so that the results, which are stated before, are applicable, let $\underline{Y}' = (y_{1,1},\ldots,y_{1,J};\ldots;y_{K,1},\ldots,y_{K,J})$ and let $\underline{Z}' = (\underline{Z}_1',\underline{Z}_2') = \underline{Y}'A$ where A is a KJ × KJ column orthogonal matrix, $\underline{Z}_1'$ is 1×K such that $Z_{1,i} = \frac{1}{J}\sum_{j=1}^{J} y_{ij}$ for $i = 1,\ldots,K$ and $\underline{Z}_2'$ is 1 × K(J-1). From the structure of Σ and the column orthogonality of A, it is easy to show that $\underline{Z}_1$ and $\underline{Z}_2$ are independent, $\underline{Z}_1$ has a normal distribution with mean vector $\underline{\theta}$, $(\theta_i = \beta_i+\delta, i = 1,\ldots,K)$ and a K×K covariance matrix $\Sigma_1 = \frac{\sigma^2}{J}[(1-\rho)I+J\rho\underline{e}\underline{e}']$, and $\underline{Z}_2$ has a normal distribution with mean vector $\underline{\xi}$ which does not depend on $\underline{\beta}$ and a diagonal covariance matrix Σ_2. Since the loss functions $L_\gamma(\underline{\beta})$ are translation invariant, an equivalent problem is to rank the parameter vector $\underline{\theta}$ with loss functions $L_\gamma(\theta_1,\ldots,\theta_K) = L_\gamma(\beta_1+\delta,\ldots,\beta_K+\delta)$. The problem is now in the form described before and the earlier results are applicable. That the density of $\underline{Z}_1$ has property M follows from Theorem 1.2.7.

4.2.11 Example. Ranking Variances in Normal Populations.

Consider independent observations Y_{ij}, $i = 1,\ldots,k$, $j = 1,\ldots,n$, where Y_{ij} is $N(\mu_i,\sigma_i^2)$ for $j = 1,\ldots,n$. The problem is to rank the unknown variances with loss functions $L_\gamma(\sigma_1^2,\ldots,\sigma_k^2)$ which depend only on the unknown variances. The sufficient statistic for this problem is $\underline{Z}' = (\underline{X}',\underline{W}')$ where $W_i = \frac{1}{n}\sum_{j=1}^{n} Y_{ij}$ and $X_i = \sum_{j=1}^{n}(Y_{ij}-W_i)^2$ for $i = 1,\ldots,k$. The problem is clearly invariant under translations of the vector $\underline{W}$

by $\underline{b} \in \mathbb{R}^k$, and any invariant decision rule will be a function of $\underline{X}$ only. But for such invariant decision functions, the earlier results are directly applicable since the density of $\underline{X}$ has property M.

Alam (1973) uses the preceding methods to construct a standard procedure φ_0 with a family of distributions having SIP for selecting a set of m < k coordinate values corresponding to the m largest components of $\underline{\theta}$. For this selection problem, we partition the set $\{1,2,...,k\}$ into two disjoint subsets γ_1 and γ_2, consisting of m and k-m elements, respectively. Some examples are given by Alam (1973) to show that the results are applicable.

Gupta and Huang (1977) study a subset selection problem (which is not symmetric in general) for the elliptically contoured density.

Now, let us consider a situation where we are interested in the subset selection problem. As usual, let the ordered parameters be $\theta_{[1]} \leq \cdots \leq \theta_{[k]}$. We are given a random observable X_i from the population π_i, $i = 1,2,...,k$. The vector of observation $\underline{x} = (x_1,...,x_k)$ of $\underline{X} = (X_1,...,X_k)$ is assumed to have density $f(\underline{x}; \underline{\theta})$. The population associated with the largest parameter $\theta_{[k]}$ is called the best and a selection of a subset containing the best population is called a correct selection (CS). In the case where two or more of the largest parameter values are equal to one of these parameters or the corresponding populations is "tagged" and called the best population.

Consider a selection of the subset $a \in G$ results in a loss

(4.2.16) $$L(\underline{\theta},a) = \sum_{i \in a} L_i(\underline{\theta})$$

where $L_i(\underline{\theta})$ is the loss whenever the ith population is selected. An additional loss of L will be imposed if a correct selection is not made.

Now assume that the values $\theta_{[1]},...,\theta_{[k]}$ are known. Under this assumption, Studden (1967) has proposed a rule to minimize the risk

(4.2.17) $$R(\underline{\theta},\delta) = \sum_{i=1}^{k} L_i(\underline{\theta}) E_{\underline{\theta}} \delta_i(\underline{X}) + L[1-P_{\underline{\theta}}(CS|\delta)].$$

Note that we let $R_1(\underline{\theta},\delta) = \sum_{i=1}^{k} L_i(\underline{\theta}) E_{\underline{\theta}} \delta_i(\underline{X})$,

(i) $L(\underline{\theta}) \equiv 1$. In this situation $R_1(\underline{\theta},\delta) = \sum_{i=1}^{k} E_{\underline{\theta}} \delta_i(\underline{X})$ is the expected size of the selected subset.

(ii) $L_i(\underline{\theta}) = 1$ if $\theta_i \neq \theta_{[k]}$,
$\qquad\quad = 0$ if $\theta_i = \theta_{[k]}$,

for $i = 1,...,k$. In this case $R_1(\underline{\theta},\delta)$ is the expected subset size excluding the best population.

Consider the k independent populations with exponential distributions and let $\theta_{[1]} = \cdots = \theta_{[k-1]} = \theta_{[k]} - \Delta$, $(\Delta > 0)$. Studden (1967) has derived two classical selection rules of Gupta (1965) and Seal (1955) as $\Delta \to \infty$ and $\Delta \to 0$, respectively.

Bickel and Yahav (1977) consider the same problem with the following loss function to derive an optimal (uniformly best) invariant procedure:

$$L(\underline{\theta},S) = \theta_{[1]} - \frac{1}{r}\sum_{j=1}^{r}\theta_{i_j} + AI_{\{\pi_{(1)} \notin S\}},$$

where $\pi_{(j)}$ is the population with mean $\theta_{[j]}$, A is a positive constant, and I_S is the indicator function of S. Note that the above loss includes the penalty for including in the subset non-best populations; it also includes a loss incurred by a "wrong selection", i.e., by not including the best population.

Moreover, suppose $\theta_1,\ldots,\theta_k$ are the unknown means. Let J be a subset of $\{1,2,\ldots,k\}$. Chernoff and Yahav (1977) consider the following loss function:

$$L(\underline{\theta},J) = (1+r)\max_{1\leq i\leq k}\theta_i - V(\underline{\theta},J),$$

where

$$V(\underline{\theta},J) = \sum_{i\in J}\theta_i/|J| + r\max_{i\in J}\theta_i,$$

$|J|$ = the number of elements in J and r is a positive constant. Note that the loss function $L(\underline{\theta},J)$ compares the value $V(\underline{\theta},J)$ with the best achieved by the experimenter who correctly guesses the best treatment. Chernoff and Yahav (1977) have made some calculations in an invariant Bayesian framework using normal distributions and normal priors with respect to the loss function $L(\underline{\theta},J)$. They use Monte Carlo simulations for the cases of k = 3 and 8 populations to show that the rule of Gupta (1965) is highly efficient.

Deely and Gupta (1968) proved that the Bayes rule for selecting a subset of k normal populations selects only one population if the loss function is a linear combination of $L_i(\underline{\theta}) = \theta_{[k]} - \theta_i$, i = 1,...,k. Goel and Rubin (1977) consider nonlinear loss functions which are based on the loss depending on the populations selected in the subset and depending on whether or not the selected subset contains the best population as follows:

$$L(\underline{\theta},s) = c|s| + [\theta_{[k]} - \theta_{\{s\}}],$$

where $\theta_{\{s\}}$ denotes $\max_{j:\ \pi_j\in S}\theta_j$ and c > 0 is to be interpreted as the relative cost of further evaluating a population versus being one unit away from the best population. A Bayes rule is obtained by Goel and Rubin (1977).

Gupta and Hsu (1978) have studied Bayes rules for selecting the best of several normal means under exchangeable normal priors and under the loss function

$$L(\underline{\theta},s) = c_1|s| + c_2 ICS(\underline{\theta},s)$$

where $ICS(\underline{\theta},s) = 0$ if $\theta_{[k]}$ is in the selected set and 1 otherwise.

The Monte Carlo studies of Gupta and Hsu indicate that the Gupta-type maximum means procedure is "quite close" to the Bayes rule in its performance.

In many practical situations, we don't know accurately the distributions which

are derived from the data. The proposed procedures may be very sensitive when the
distributions considered are different in form such as the normal and the double
exponential families. Also, sometimes we don't know exactly the numerical values of
the observations, such as the failure time for some systems (structures) in reliability problems. In these cases we may have numerical values of ranks available for
use. Hollander and Wolfe (1973) point out that although, at first glance, most nonparametric procedures based on ranks seem to sacrifice too much of the basic information in the samples, theoretical investigations have shown that this is not the
case. More often than not, the nonparametric procedures are only slightly less
efficient than their normal theory competitors when the underlying populations are
normal and they can be mildly more efficient than these competitors when the underlying populations are not normal. As we know, the developments for the theory of
rank tests are based on the criteria of Neyman-Pearson approach (see Hajek and Sidák
(1967)), or based on various reasonable ways to derive procedures and then to study
their efficiency. Both of the preceding approaches may have certain deficiencies.
An alternative way is to derive procedures based on the experimenter's goal in practical applications. This is especially true in multiple decision problems. Gupta,
Huang and Nagel (1979) have studied a procedure for which the probability of a correct
selection is "as sensitive as possible" when the parameters change from all equal to
unequal in some neighborhood; these are locally optimal procedures based on ranks.
In some special cases, a locally optimal rule can be derived as the procedure based
on Wilcoxon rank-sum statistic. The P*-condition can be guaranteed in this case
and various kinds of efficiency properties have been studied extensively in literature (see Hajek and Sidák (1967)).

Now we discuss some subset selection procedures based on ranks. From each of
the populations π_i, $i = 1,2,\ldots,k$, we take n independent observations $X_{i1},\ldots,X_{in}$.
Let R_{ij} denote the rank of X_{ij} in the pooled sample of the $N = kn$ observations
$(X_{11},\ldots,X_{1n};\ldots; X_{k1},\ldots,X_{kn})$.

4.2.12 <u>Definition</u>. A rank configuration is an N-tuple $\Delta = (\Delta_1,\ldots,\Delta_N)$, $\Delta_i \in \{1,2,\ldots,k\}$, where $\Delta_i = j$ indicates that the ith smallest observation in the pooled
sample comes from π_j, i.e. there exists an ℓ such that $R_{j\ell} = i$ holds.

Let $C = \{\Delta\}$ denote the set of all rank configurations for k and n which are kept
fixed in these considerations. $\Delta_{\underline{x}}$ denotes the rank configuration of $\underline{x} = (x_1,\ldots,x_N)$.
For a fixed Δ, let $\mathcal{X}_\Delta = \{\underline{x} \in \mathcal{X} | \Delta_{\underline{x}} = \Delta\}$, where $\mathcal{X} = \{\underline{x} | \underline{x} = (x_1,\ldots,x_N)\}$. The decision
space $\mathcal{D}$ consists of the 2^k subsets of the set $\{1,2,\ldots,k\}$:

$$\mathcal{D} = \{d | d \subset \{1,2,\ldots,k\}\}.$$

A decision is the selection of a subset of the k populations. The fact that $i \in d$
indicates that π_i is included in the selected subset if decision d is made.

Let $\delta(\Delta,d)$ denote the conditional probability that decision d is made if the
rank configuration Δ is observed. We restrict considerations to the functions

$p_i(\Delta) = \sum_{d \ni i} \delta(\Delta,d)$, $i = 1,2,\ldots,k$, as follows.

Let the distribution of π_i be given by a density function $f(x,\theta_i)$ from a one parameter family with the θ_i's belonging to some interval, which, without loss of generality, can be assumed to contain zero. Let $\Theta = \{\underline{\theta}|\underline{\theta} = (\theta_1,\ldots,\theta_k)\}$. Furthermore, let the family $f(x,\theta)$ have the following properties:

<u>Condition A</u>.
(i) $f(x,\theta)$ is absolutely continuous in θ for almost every x;
(ii) the limit
$$\dot{f}(x,0) = \lim_{\theta \to 0} \frac{1}{\theta}[f(x,\theta)-f(x,0)]$$
exists for almost every x;
(iii)
$$\lim_{\theta \to 0} \int_{-\infty}^{\infty} |\dot{f}(x,\theta)|dx = \int_{-\infty}^{\infty} |\dot{f}(x,0)|dx < \infty$$

holds, with $\dot{f}(x,\theta)$ denoting the partial derivative with respect to θ. Note that the existence of $\dot{f}(x,\theta)$ for almost every θ is ensured at every point x such that $f(x,\theta)$ is absolutely continuous in θ. This, however, does not make the condition (ii) superfluous (see Hajek and Sidák (1967)).

Gupta, Huang and Nagel (1979) have proved that for any i, $(1 \leq i \leq k)$, if

$$p_i(\Delta) = \begin{cases} 1 & \text{if } A_i(\Delta) > c, \\ \rho_i & \text{if } A_i(\Delta) = c, \\ 0 & \text{if } A_i(\Delta) < c, \end{cases}$$

where

$$A_i(\Delta) = A \sum_{\substack{j \\ \Delta_j = i}} \int_{-\infty}^{x_N} \int_{-\infty}^{x_2} \cdots \int_{-\infty}^{\infty} \dot{f}(x_j,0) \prod_{\substack{\ell=1 \\ \ell \neq j}}^{N} f(x_\ell,0) dx_1 \ldots dx_N,$$

(A is a constant) satisfies $\inf_{\theta_1 = \ldots = \theta_k} P_{\underline{\theta}}(CS|\delta,\Delta) = P^*$, then it makes $P_{\underline{\theta}}(CS|\delta,\Delta)$ as large as possible in the neighborhood $0 < |\theta_i| < \epsilon$, $1 \leq i \leq k$, for given $\epsilon > 0$ among all invariant rules based on ranks satisfying $\inf_{\theta_1 = \ldots = \theta_k} P_{\underline{\theta}}(CS|\delta,\Lambda)=P^*$. The constant ρ_i and c are determined by
$$\sum_{A_i(\Delta)>c} P_{\underline{\theta}_0}(\Delta) + \rho_i \sum_{A_i(\Delta)=c} P_{\underline{\theta}}(\Delta) = P^*.$$

Note that this locally optimal rule is based on weighted rank sums using the scores
$$B_i = \int_{-\infty}^{\infty} u^{i-1}(1-u)^{N-i}\phi(u,f)du,$$

where

$$\varphi(u,f) = \frac{\dot{f}(F^{-1}(u,0),0)}{f(F^{-1}(u,0),0)},$$

and F is the cdf of the density f.

If f has the logistic density

$$f(x,\theta) = e^{-(x-\theta)}/[1+e^{-(x-\theta)}]^2$$

then $\varphi(u,f) = 2u-1$ which leads to equally spaced scores: $B_i = a+ib$ where the actual values of a and $b > 0$ are irrelevant. Hence the locally optimal rule is as follows:

$$\text{Select } \pi_i \text{ iff } \sum_{j=1}^{n} R_{ij} \geq c.$$

Note that Nagel (1970) has shown that the rules of this type are 'just' provided that B_i's are non-decreasing in i, which, for location parameters, is true if and only if f(x) is strongly unimodal. It follows from Nagel (1970) that

$$\inf_{\underline{\theta}\in\Omega} P_{\underline{\theta}}(CS|\delta,\Delta) = \inf_{\underline{\theta}\in\Omega_0} P_{\underline{\theta}}(CS|\delta,\Delta)$$

for a 'just' selection rule δ. Thus the selection rule R satisfies the P*-condition.

References

[1] Alam, K. (1973). On a multiple decision rule. <u>Ann. Statist.</u> 1, 750-755.

[2] Bahadur, R. R. and Goodman, L. A. (1952). Impartial decision rules and sufficient statistics. <u>Ann. Math. Statist.</u> 23, 553-562.

[3] Bickel, P. J. and Yahav, J. A. (1977). On selecting a set of good populations. <u>Statistical Decision Theory and Related Topics II</u> (Eds. Gupta, S. S. and Moore, D. S.), Academic Press, New York, 37-55.

[4] Chernoff, H. and Yahav, J. (1977). A subset selection problem employing a new criterion. <u>Statistical Decision Theory and Related Topics II</u> (Gupta, S. S. and Moore, D. S.), Academic Press, New York, 93-119.

[5] Deely, J. J. and Gupta, S. S. (1968). On the properties of subset selection procedures. <u>Sankhyā Ser.</u> A 30, 37-50.

[6] Eaton, M. L. (1967). Some optimum properties of ranking procedures. <u>Ann. Math. Statist.</u> 38, 124-137.

[7] Ferguson, T. S. (1967). <u>Mathematical Statistics: A Decision Theoretic Approach</u>. Academic Press, New York.

[8] Goel, P. K. and Rubin, H. (1977). On selecting a subset containing the best population - A Bayesian approach. <u>Ann. Statist.</u> 5, 969-983.

[9] Gupta, S. S. (1965). On some multiple decision (selection and ranking) rules. <u>Technometrics</u> 7, 225-245.

[10] Gupta, S. S. and Hsu, J. C. (1978). On the performance of some subset selection procedures. <u>Comm. Statist.-Simula. Computa.</u> B7, 561-591.

[11] Gupta, S. S. and Huang, D. Y. (1977). Some multiple decision problems in analysis of variance. <u>Comm. Statist.-Theor. Meth.</u> A6 (11), 1035-1054.

[12] Gupta, S. S., Huang, D. Y. and Nagel, K. (1979). Locally optimal subset selection procedures based on ranks. Optimizing Methods in Statistics (Ed. J. S. Rustagi), Academic Press, New York, 251-260.

[13] Hajek, J. and Sidák, Z. (1967). Theory of Rank Tests. Academic Press, New York.

[14] Hollander, M. and Wolfe, D. A. (1973). Nonparametric Statistical Methods. John Wiley and Sons, New York.

[15] Lehmann, E. L. (1966). On a theorem of Bahadur and Goodman. Ann. Math. Statist. 37, 1-6.

[16] Nagel, K. (1970). On subset selection rules with certain optimality properties. Ph.D. Thesis (Mimeo. Ser. No. 222). Dept. of Statist., Purdue Univ., W. Lafayette, IN 47907.

[17] Seal, K. C. (1955). On a class of decision procedures for ranking means of normal populations. Ann. Math. Statist. 36, 387-397.

[18] Studden, W. J. (1967). On selecting a subset of k populations containing the best. Ann. Math. Statist. 38, 1072-1078.

CHAPTER 5
ROBUST SELECTION PROCEDURES: MOST ECONOMICAL MULTIPLE DECISION RULES

5.1 Introduction

In Section 5.2, we deal with two important aspects of multiple decision rules for the robustness of rules to determine the optimal sample size. The most economic rule, roughly defined to be a rule with least sample size to meet the goal, is described later. Robustness is, of course, a very important consideration in practical applications.

5.2. Robust Selection Rules

An experimenter is asked to make decision regarding the k populations and must decide how large a sample he should take to decide this question. Taking a large sample decreases the probability of an incorrect decision, but at the same time increases the cost of sampling. It seems reasonable that the "optimum" sample size should depend both on the cost of sampling and the amount of use to be made of the decision. Loss functions should then be set up which take into consideration the amount of use to be made of the result, the cost of making a wrong decision and the cost of sampling.

In the consideration of robustness of decision procedures, we are concerned with the following concept of optimality (see Schäl (1979)).

Let $\mathcal{D}$ be the decision space and let Θ be the parameter space. The total expected cost $R(\theta,d)$ depends not only on the decision $d \in \mathcal{D}$ but also on the parameter $\theta \in \Theta$ and is, therefore, called the risk function. $\mathcal{P}(\Theta)$ stands for the set of all probability distributions on Θ. Let Λ be a subset of $\mathcal{D}$ and let Γ be a subset of $\mathcal{P}(\Theta)$. Then $d^* \in \Lambda$ is said to be Γ-optimal in Λ - we write $d^* \in \Lambda^*(\Gamma)$ - if

$$\sup_{\mu \in \Gamma} R(\mu, d^*) = \inf_{d \in \Lambda} \sup_{\mu \in \Gamma} R(\mu, d).$$

Note that a Γ-optimal decision rule in Λ is also called Γ-minimax in Λ. Obviously for any given $\mu \in \mathcal{P}(\Theta)$, we have $d^* \in \mathcal{D}^*(\{\mu\})$, if and only if d^* is a Bayes rule against the a priori distribution μ, i.e.

$$R(\mu, d^*) = \inf_{d \in \mathcal{D}} R(\mu, d).$$

And $d^* \in \mathcal{D}^*(\mathcal{P}(\Theta))$ if and only if d^* is a minimax rule, i.e.,

$$\sup_{\theta \in \Theta} R(\theta, d^*) = \inf_{d \in \mathcal{D}} \sup_{\theta \in \Theta} R(\theta, d),$$

where use is made of

$$\sup_{\theta \in \Theta} R(\theta, d) = \sup_{\mu \in \mathcal{P}(\Theta)} R(\mu, d), \quad d \in \mathcal{D}.$$

Further, upon defining for any $0 \leq \rho_0 \leq 1$, $\mu_0 \in \mathcal{P}(\Theta)$, $\Gamma = \{\rho_0 \mu_0 + (1-\rho_0)\nu, \nu \in \mathcal{P}(\Theta)\}$, we obtain the optimality criterion of Hodges and Lehmann (1952) where

$$\sup_{\mu \in \Gamma} R(\mu, d) = \rho_0 R(\mu_0, d) + (1-\rho_0) \sup_{\theta \in \Theta} R(\theta, d).$$

If $\Theta = \cup \Theta_i$ is a measurable partition and we choose, for some $p_i \geq 0$ with $\Sigma p_i = 1$, $\Gamma = \{\mu \in P(\Theta); \mu(\Theta_i) = p_i\}$, we obtain the optimality criterion of Menges (1966) where

$$\sup_{\mu \in \Gamma} R(\mu,d) = \Sigma p_i \sup_{\theta \in \Theta_i} R(\theta,d).$$

Let there be k (≥ 2) independent populations with continuous distribution functions $F(x_1-\theta_1)$, $F(x_2-\theta_2),\ldots,F(x_k-\theta_k)$, where the location parameters θ_i's are unknown. Let $X_{i1},\ldots,X_{in}$ denote n independent observations from the ith population and define $\bar{X}_i = \frac{1}{n}\sum_{j=1}^{n} X_{ij}$, $1 \leq i \leq k$. Let $\delta_i(\underline{x})$ denote the probability of selecting the ith population. Let Λ be the set of all usual procedures defined as follows:

$$\delta_i(\underline{x}) = \begin{cases} 1 & \text{if } \bar{x}_i \geq \max_{1 \leq j \leq k} \bar{x}_j, \\ 0 & \text{if } \bar{x}_i < \max_{1 \leq j \leq k} \bar{x}_j. \end{cases}$$

Let $\Theta = \{\underline{\theta}|\underline{\theta} = (\theta_1,\ldots,\theta_k)\}$ and $\Theta_i = \{\underline{\theta}|\theta_i \geq \max_{j \neq i} \theta_j + \Delta\}$, $i = 1,2,\ldots,k$, and Δ is a given positive constant. Then the partition of Θ is $\Theta = \Theta_0 \cup \Theta_1 \cup \cdots \cup \Theta_k$, where the Θ_0 is that part of Θ usually called indifference zone.

For $\underline{\theta} \in \Theta_i$, $1 \leq i \leq k$, define $L^{(r)}(\underline{\theta},\delta_j) = 0$ for all $r \neq i$, $L^{(i)}(\underline{\theta},\delta_j) = c'(\theta_i-\theta_j)\delta_j$, $j = 1,2,\ldots,k$, where $L^{(i)}(\underline{\theta},\delta_j)$ represents the loss for $\underline{\theta} \in \Theta_i$ when the jth population is selected, c' being a positive constant. For $\underline{\theta} \in \Omega_0$, the loss is zero.

The probability of making a wrong decision can be decreased by increasing the size of the experiment on which the decision is to be based, but this increases the cost of experimentation, which must also be considered. It will be assumed here that the cost of performing an experiment involving n observations from each population is cnk, where c is a positive constant. Let $\rho \in P(\Theta)$. Then the risk function, or the expected loss, with experimentation costs included is

(5.2.1) $\quad \gamma_n(\rho) = cnk + c' \sum_{i=1}^{k} \sum_{j=1}^{k} \int_{\Theta_i} \int_{\mathbb{R}^k} (\theta_i-\theta_j)\delta_j(\underline{x}) dF_{\underline{\theta}}(\underline{x}) d\rho(\underline{\theta}).$

Assume that some partial information is available in the selection problem, so that we are able to specify $\pi_i = P\{\underline{\theta} \in \Theta_i\}$, $\sum_{i=0}^{k} \pi_i = 1$. Define

$$\Gamma = \{\rho(\underline{\theta})|\int_{\Theta_i} d\rho(\underline{\theta}) = \pi_i, \quad i = 0,1,2,\ldots,k\}.$$

In case of normal populations, Gupta and Huang (1977) have obtained some most economical selection rules for the risk (5.2.1) and the prior information Γ. Instead of these considerations, we shall consider the criterion based on the probability of correct decision as follows.

Bechhofer (1954) has considered a single sample multiple decision procedure for selecting, among a group of normal populations with common known variances, that population with the largest mean. Bechhofer and Sobel (1954) have described a procedure for selecting the normal population with the smallest variance. Several other analogous problems have also been considered. They suggest, with only intuitive justification, selecting the population with the largest (smallest) sample mean (variance), and give tables for finding the minimum sample size (assumed equal for all populations) which will guarantee a correct decision with a prescribed probability when the extreme population parameter is sufficiently distinct from the others. Hall (1959) gives justification for a wide class of such procedures, proving that no other equal sample size rule can meet this guarantee with a smaller (fixed) sample size; that is, such rules are most economical as defined below.

We are concerned with a sequence $X_1, X_2, \ldots,$ of real-or vector-valued, independent, and identically distributed random variables, each having a density function f, belonging to some specified class $\mathcal{F}$, w.r.t. a σ-finite measure μ.

The decision problem is to formulate a rule for choosing a non-negative integer n (completely non-random), and, after taking an observation
$$\underline{x} = (x_1, \ldots, x_n)$$
on $\underline{X} = (X_1, \ldots, X_n)$, for choosing one of m possible alternative actions $a_1, \ldots, a_m$. A multiple decision rule for choosing among $a_1, \ldots, a_m$ on the basis of $\underline{x}$ is defined by an ordered set of non-negative, real-valued, measurable functions $\varphi(\underline{x}) = (\varphi_1(\underline{x}), \ldots, \varphi_m(\underline{x}))$ on the space $\mathcal{X}$ of $\underline{x}$ such that $\sum_{i=1}^{k} \varphi_i(\underline{x}) = 1$ for all $\underline{x}$. Then a_i is chosen with probability $\varphi_i(\underline{x})$ when $\underline{x}$ is observed. For non-randomized decision rules, the φ_i's are characteristic functions of mutually exclusive and exhaustive "acceptance" regions $R_1, \ldots, R_m$ in $\mathcal{X}$, where a_i is accepted if $\underline{x} \in R_i$.

A subscript or superscript n denotes the corresponding sample size; $f^n(\underline{x})$ and μ^n denote the joint density and product measure, respectively.

We suppose throughout in this section that $\mathcal{F}$ consists of a finite number, say ℓ, of elements $f_1, \ldots, f_\ell$; we say that the corresponding decision problem is one of "simple discrimination" and a decision rule is for discriminating among $f_1, \ldots, f_\ell$. Here, if μ is non-atomic, only non-randomized decision rules need to considered (see Hall (1958)).

A decision rule $d = d_n$ is characterized by the functions
$$p_{ij}(d) = P(d \text{ chooses } a_j | f_i) = \int_{\mathcal{X}} \varphi_j(\underline{x}) f_i^n(\underline{x}) d\mu^n,$$
for $i = 1, \ldots, \ell; j = 1, \ldots, m$.

We consider two different criteria for choosing a decision rule for simple discrimination. The first assumes that $\ell = m$ and that the action a_i is to be preferred when f_i is true. Denote $p_{ii}(d) = p_i(d) = 1 - q_i(d)$, so that p_i is the probability of a "correct" decision and q_i the probability of an "incorrect" decision when f_i is true.

5.2.1 <u>Definition</u>. Let $\underline{\alpha} = (\alpha_1,\ldots,\alpha_m)$ be a given vector of positive constants each less than one. A decision rule d_N, based on a sample of size N, is said to be a most economical m-decision rule relative to the vector $\underline{\alpha}$ for discriminating among $f_1,\ldots,f_m$ if it satisfies

(5.2.2) $\qquad\qquad\qquad p_i(d) \geq \alpha_i \qquad (i = 1,\ldots,m),$

and if N is the least integer n for which (5.2.2) may be satisfied by some m-decision rule d_n based on a sample of size n. N is said to be the most economical sample size.

We now no longer require that $\ell = m$, but suppose that corresponding to each f_i one or more of the alternatives a_j is preferable or "correct", when f_i is true.

5.2.2 <u>Definition</u>. Let $\beta = (\beta_{ij})$ be a given $\ell \times m$ matrix of positive constants such that for every i, j pair for which a_j is a correct decision when f_i is true $\beta_{ij} = 1$. A decision rule d_N, based on a sample of size N, is said to be most economical m-decision rule relative to the matrix β for discriminating among $f_1,\ldots,f_\ell$ if it satisfies

(5.2.3) $\qquad\qquad\qquad p_{ij}(d) \leq \beta_{ij} \qquad (i = 1,\ldots,\ell;\ j = 1,\ldots,m)$

and if N is the least integer n for which (5.2.3) may be satisfied by some m-decision rule d_N based on a sample of size n. N is said to be the most economical sample size.

Hall (1958) points out that if $\ell = m$ and a_i is preferred when f_i is true, then a most economical decision rule relative to β also controls the probabilities of correct decisions if $\sum_{j \neq i} \beta_{ij} < 1$ for all i. If $\ell = m = 2$, both (5.2.2) and (5.2.3) reduce to upper bounds on the probabilities of two kinds of error, and Definitions 5.2.2 and 5.2.3 provide a most economical 2-decision rule as the one with minimum sample size subject to these bounds.

The proof of the most economical character of decision rules is achieved by proving their minimax character when a suitable loss function is introduced. Bahadur and Goodman (1952) have considered a class of multiple decision rules they have called impartial (invariant under permutations of the populations). Their results are applicable to such problems of choosing the best population and imply that Bechhofer and Sobel's rules are minimax rules (in fact, uniformly minimax risk rules) among the class of impartial decision rules. Hall (1959) has shown that impartiality is no restriction when looking for minimax rules, as is well-known to be the case for certain other kinds of invariance. Eaton (1967) has extended some results of Hall (1958, 1959), concerning most economical decision rules, to include the class of densities with property M.

Now we consider the case where the form of the distribution is not known exactly. We consider the robust selection procedures as follows.

Let X_{ij} $(j = 1,\ldots,n;\ i = 1,\ldots,k)$ be independent observations from k populations with respective distribution functions (cdf's) $F(x-\theta_i)$, and let $\theta_{[1]} \leq \cdots \leq \theta_{[k]}$ denote the ordered θ_i's. We consider the problem of selecting the "best" population, namely, the one with the largest location parameter θ; the methods are readily extended to other ranking and selection goals, as introduced by Bechhofer (1954) and treated recently by Ghosh (1973); see also Gibbons, Olkin and Sobel (1977).

Let $\Delta(>0)$ and $P^*(>\frac{1}{k})$ be specified and write $S = S_\Delta$ for the subset of the parameter space R^k of $\underline{\theta} = (\theta_1,\ldots,\theta_k)$ where $\theta_{[1]} \leq \cdots \leq \theta_{[k]}$ and $\theta_{[k]} \geq \theta_{[k-1]} + \Delta$. The problem is to choose a value, say N, for the (common) sample size n and a selection procedure for choosing the best population which assures a PCS (probability of correct selection) of at least P^* whenever $\underline{\theta} \in S$. When F is Φ - and the errors are thus standard normal - Bechhofer (1954) showed that, by choosing $N = d^2/\Delta^2$ (or $\geq$) with $d = P^{-1}(P^*)$ and

(5.2.4) $$P(d) \equiv \int \Phi^{k-1}(x+d)d\Phi(x),$$

the procedure which selects the population with the largest sample mean meets this goal:

$$PCS_N(\Phi,\underline{\theta}) \geq P^* \text{ when } \underline{\theta} \in S_\Delta,$$

with equality at a least favorable configuration (ℓfc) where all but one of the populations have the same θ-value and the remaining population has a θ-value which is Δ units larger. Hall (1959) showed this procedure to be most economical, in the sense that no competing procedure could meet the goal with a smaller sample size.

Dalal and Hall (1979) continue to mention that other authors have introduced various competing procedures, especially procedures based on various nonparametric statistics (e.g. Lehmann (1963), Randles (1970) and Ghosh (1973); for other references see Ghosh (1973)). By selecting as the best population the one with the largest location parameter estimator (using a nonparametric estimate of location), the procedures are not so sensitive to the assumption of normal errors. However, they are not fully nonparametric, even asymptotically, in that the rule for choosing N is $N \sim d^2 \tau^2/\Delta^2$ (as $\Delta \downarrow 0$) where τ^2/n is the (asymptotic) variance of the location estimator T_n used by the selection rule. Since $\tau = \tau(F)$, so now assumptions about F are needed in order to determine N. For example, one may use the Hodges-Lehmann location estimator and then $\tau^2 = (2\sqrt{3} \int f^2)^{-2}$, assuming F has density f (see Ghosh (1973)); specifically, $\tau^2(\Phi) = \frac{\pi}{3}$. The ratio of N's for procedures based on a location estimator T_n and on the sample mean, respectively, is asymptotically $\tau^2(\Phi)$ - assuming both N's were chosen to meet the PCS requirement at $F = \Phi$. Such ARE's are discussed by Ghosh.

Asymptotically, the choice of

$$N = N(\Delta) = d^2 \tau^2(\Phi)/\Delta^2 + o(\Delta^{-2})$$

works for other F's, with the same or smaller $\tau(F)$ in that

$$\lim_{\Delta \downarrow 0} \inf_{\underline{\theta} \in S_\Delta} PCS_{N(\Delta)}(F,\underline{\theta}) \geq P^*$$

for each such F. Dalal and Hall (1979) point out that one cannot claim that, for small $\Delta = \Delta(\epsilon)$,

$$\inf_{\underline{\theta}\in S} PCS_{N(\Delta)}(F,\underline{\theta}) > P^*-\epsilon$$

for each such F; for this, one would need the $o(\Delta^{-2})$ term above to be uniform in F, thus allowing the insertion of an "inf over F" in front of the PCS. Thus, the asymptotic formulation of the problem is: find a sequence of selection procedures (one for each sample size) and a sample size formula $N = N(\Delta)$ such that

(5.2.5) $$\lim_{\Delta\downarrow 0} \inf_{F\in\mathcal{F}, \underline{\theta}\in S_\Delta} PCS_{N(\Delta)}(F,\underline{\theta}) \geq P^*(> \frac{1}{k}).$$

A procedure satisfying (5.2.5), for some suitable $\mathcal{F}$ containing Φ, is said to be asymptotically robust at $\Phi \in \mathcal{F}$ - since we have P^*-protection at F's near Φ. If every other procedure satisfying (5.2.5), but with $N' = N'(\Delta)$, has the property

$$\lim_{\Delta\downarrow 0} \inf [N'(\Delta)/N(\Delta)] \geq 1,$$

then the procedure is asymptotically most economical for $\mathcal{F}$ as well - N is minimal.

Thus the approach of Dalal and Hall (1979) differs in that they set out to choose a suitable family $\mathcal{F}$ of possible error distributions and determine N (minimally) so as to meet the PCS goal for every $F \in \mathcal{F}$ - specifically for a least favorable F^0 in $\mathcal{F}$. This continues the same minimax spirit of these selection procedures: they not only guard against a least favorable configuration of location parameter values, but against a least favorable error distribution as well - the roles of θ and F are treated in the same way. For this development, the "contamination" or "gross error" model approach of Huber (1964) has been studied by Dalal and Hall (1979) as follows.

Let $\mathcal{F} = \{F | F = (1-\gamma)\Phi+\gamma H, H$ is an arbitrary df symmetric at $0\}$, for specified $\gamma \in [0,1)$. This is a (symmetric) "contamination neighborhood" of the ideal model Φ, allowing for a proportion γ or less of "gross errors" in the model. A selection procedure based on Huber's M-estimate T^0 (see Dalal and Hall (1964)), with sample size chosen to meet the PCS goal (asymptotically) at F^0, meets the goal for every $F \in \mathcal{F}^0$, an equicontinuous (at the censoring point) subset of $\mathcal{F}$, and the appropriate asymptotic sample size N is $d^2\sigma_0^2/\Delta^2$ ($\sigma_0^2 \equiv \sigma(F^0)^2$); moreover, this N is minimal. We conclude therefore that the procedure is (asymptotically) a most-economical robust selection procedure. The ratio of sample sizes for this procedure relative to Bechhofer's $\bar{X}$-procedure is thus σ_0^2. This ratio differs conceptually from the ARE (see Ghosh (1973); it measures how much the sample size needs to be increased to expand the P^*-protection from Φ to $\mathcal{F}^0$, whereas ARE compares two procedures both evaluated at Φ. The PCS of Dalal and Hall (1979) procedure when F is actually Φ is at least $P(d\sigma_0)$, which is greater than P^*, while the PCS for the $\bar{X}$-procedure when $F \neq \Phi$ but $\in \mathcal{F}^0$ may be smaller than P^*.

References

[1] Bahadur, R. R. and Goodman, L. A. (1952). Impartial decision rules and sufficient statistics. Ann. Math. Statist. 23, 553-562.

[2] Bechhofer, R. E. (1954). A single sample multiple decision procedure for ranking means of normal populations with known variances. Ann. Math. Statist. 25, 16-39.

[3] Bechhofer, R. E. and Sobel, M. (1954). A single sample multiple decision procedure for ranking variances of normal populations. Ann. Math. Statist. 25, 273-289.

[4] Dalal, S. R. and Hall, W. J. (1979). Most economical robust selection procedures for location parameters. Ann. Statist. 7, 1321-1328.

[5] Eaton, M. L. (1967). Some optimum properties of ranking procedures. Ann. Math. Statist. 38, 124-137.

[6] Ghosh, M. (1973). Nonparametric selection procedures for symmetric location parameter populations. Ann. Statist. 1, 773-779.

[7] Gibbons, J. D., Olkin, I. and Sobel, M. (1977). Selecting and Ordering Populations: A New Statistical Methodology. Wiley, New York.

[8] Gupta, S. S. and Huang, D. Y. (1977). On some optimal sampling procedures for selection problems. Theory and Applications of Reliability with Emphasis on Bayesian and Nonparametric Methods (Ed. Tsokos, C. P. and Shimi, I. N.), Academic Press, New York, 495-505.

[9] Hall, W. J. (1958). Most economical multiple decision rules. Ann. Math. Statist. 29, 1079-1094.

[10] Hall, W. J. (1959). The most economical character of some Bechhofer and Sobel decision rules. Ann. Math. Statist. 30, 964-969.

[11] Hodges, J. L. Jr. and Lehmann, E. L. (1952). The use of previous experience in reaching statistical decisions. Ann. Math. Statist. 23, 396-407.

[12] Huber, P. J. (1964). Robust estimate of a location parameter. Ann. Math. Statist. 35, 73-101.

[13] Lehmann, E. L. (1963). A class of selection procedures based on ranks. Math. Ann. 150, 268-275.

[14] Menges, G. (1966). On the "Bayesification" of the minimax principle. Unternehmensforschung 10, 81-91.

[15] Randles, R. H. (1970). Some robust selection procedures. Ann. Math. Statist. 41, 1640-1645.

[16] Schäl, M. (1979). On dynamic programming and statistical decision theory. Ann. Statist. 7, 432-445.

CHAPTER 6
MULTIPLE DECISION PROCEDURES BASED ON TESTS

6.1 Introduction

In this chapter, we shall study multiple decision procedures in terms of hypotheses-testing problems. First, we discuss the conditional confidence approach of Kiefer which can be used to improve Neyman-Pearson (NP) formulation. This is done in Section 6.2. Using this approach, we describe conditional selection procedures and their relation with classical selection rules. Later, we discuss the theory of multiple comparisons for some appropriate alternative hypotheses. In Section 6.3, we consider an optimal criterion to improve the power of the individual test. Using this approach, we derive selection rules based on tests. Multiple range tests are studied in Section 6.4. A discussion of the multistage comparison procedures is provided in Section 6.5.

6.2 Conditional Confidence Approach

Conditional inference is of course an old idea, applied in many settings beginning (at least) with Fisher, who introduced the notion of an ancillary statistic partly as a basic for conditioning. Kiefer (1977) points out that there is a large body of literature on this topic. His emphasis, however, is on constructing procedures with conditional confidence coefficient that is highly variable, and on giving a coherent framework for construction and theoretical assessment of a wide variety of conditional confidence procedures under many statistical settings. Despite the extensive work on conditioning of the earlier authors, it had not been accorded such a comprehensive treatment in the literature till Kiefer's contributions. Again although one knows that conditioning is an old and commonly employed tool in statistical inference; no comprehensive framework for comparing the properties of different conditional procedures had been studied until recently by Kiefer (1975, 1976, 1977). Brownie and Kiefer (1977) have discussed the problem of testing two simple hypotheses using the conditional procedure approach. Their research gives a highly variable data-dependent measure of conclusiveness in the conclusion inferred from the experiment, with frequentist interpretability of that measure. The departure from the usual NP approach can be explained in the simple context (see Brownie and Kiefer (1977)) of testing H_0: X has density f_i, with possible decision d_0 and d_1, where d_i is the decision to accept H_i. The ideas are as follows:

(i) partition the sample space $\mathcal{X}$ into a family of subsets $C = \{C_i^b, b \in B, i = 0,1\}$, where B is a set of labels (one element in the NP case);

(ii) when the sample X falls in C_i^b, we state a conclusion (Γ^b, d_i) that "H_i is true, with (conditional) confidence Γ^b";

(iii) the Γ^b (like the level or an ordinary confidence coefficient) has a frequency interpretation, but it is now a conditional one: writing $C^b = C_0^b \cup C_1^b$, we shall have, for i = 0 and 1,

(6.2.5) $P_i\{$making the correct decision $d_i | C^b\}$
$= \Gamma^b$, w.p.1.

The frequency interpretation is clear: Suppose for simplicity that B is discrete, so that it is reasonable to assume that $P_i\{C^b\} > 0$ for all b and i. Over many experiments in which an experimenter has obtained a value $\Gamma^b = 0.95$, he will very likely be making a correct assertion in approximately 95% of those experiments.

To see the relation to the NP theory, suppose B has a single element b and that $\mathcal{X}$ is partitioned into $\mathcal{X} = C^b = C_0^b \cup C_1^b$, where C_1^b is the critical region chosen to make both probabilities of error equal to $1-\Gamma^b$. In this case we see that (6.2.5) specializes to express this probability structure in the NP setting.

Kiefer (1977) has given the following example to illustrate these ideas described above.

6.2.1 Example. Suppose we observe a normally distributed random variable X with mean θ and unit variance, and must decide between the two simple hypotheses $H_0: \theta = -1$ and $H_1: \theta = 1$. The unconditional NP procedure has a single element in B, say b, and $C_0^b = (-\infty, 0)$, while $C_1^b = [0, \infty)$. The set C^b is the entire real line; thus the probabilities in (6.2.5) are unconditional. We see that the (unconditional) confidence coefficient Γ^b is simply $0.84 = 1-\alpha = 1-\beta$, which is the complement of the common error probability value. As an example of a partition of the type described in (i) above, we might let B consist of two elements, $B = \{w, s\}$, this labeling being used to suggest weak and strong feelings of conclusiveness about the decision. A simple partition C is given by choosing a value $c > 0$ and setting

(6.2.6)
$$C_0^s = (-\infty, -c], \quad C_1^s = [c, \infty),$$
$$C_0^w = (-c, 0), \quad C_1^w = [0, c).$$

Thus the set C^s, where a strongly conclusive decision is made, is $\{x: |x| \geq c\}$, while in $C^w = \{x: |x| < c\}$ we make a weakly conclusive decision. For example, if we choose $c = 1$, the conditional confidence coefficient of (6.2.5) is easily computed in terms of the standard normal distribution function Φ to be

(6.2.7)
$$\Gamma^w = P_1\{C_1^w | C^w\} = [\Phi(0) - \Phi(-1)] / [\Phi(0) - \Phi(-2)]$$
$$= 0.71,$$
$$\Gamma^s = P_1\{C_1^s | C^s\} = \Phi(0) / [\Phi(0) + \Phi(-2)] = 0.96.$$

Note that the same values of Γ^w and Γ^s are obtained if the computation is made under P_0. Consequently, if we use this partition with $c = 1$ and we obtain the observed value $X = 0.5$ given in the illustration, we state "H_1 is true with conditional confidence 0.71"; while if $X = 5$, we state "H_1 is true with conditional confidence 0.96". We have thus achieved our goal of stating a larger measure of conclusiveness in the latter case, than in the former.

Composite Hypotheses and Multidecision Example: Selection and Ranking

Kiefer (1977) points out that the problems of selection and ranking offer a simple and natural setting for the above approach and some examples will illustrate this briefly with a few of the simplest cases as follows:

6.2.2 Example. Two normal populations. We first consider the conditional confidence parallel of the Bechhofer (1954) formulation, in the following part (a), and of the Gupta (1956) approach in (b).

For simplicity, assume the variances and sample sizes n for the two populations are equal. After a scale transformation and reduction by sufficiency, the setup becomes $X' = (X_1', X_2')$ where the X_i' are independent and normal with means θ_i and variances n^{-1}. A value $\Delta^* > 0$ is specified, the difference in means that is worth detecting. The decision d_i', (asserting θ_i is larger) is correct if $\theta_i - \theta_{3-i} > -\Delta^*, i=1,2$; thus we are indifferent to differences of no more than Δ^* in $|\theta_1 - \theta_2|$. The main interest is of course in (θ_1, θ_2) for which only one d_i' is correct, outside the indifference region. A further reduction by invariance transforms the problem to $X \sim N(0,1)$ and decisions d_i, where $\Omega(d_1) = \{\theta: \theta > -\bar{\Delta}\}$ and $\Omega(d_0) = \{\theta: \theta < \bar{\Delta}\}$; here X, θ, $\bar{\Delta}$, are $(n/2)^{\frac{1}{2}}$ times $X_1 - X_2$, $\theta_1 - \theta_2$, Δ^*, respectively. The operating characteristic is of interest when $|\theta| \geq \bar{\Delta}$. Thus the problem is reduced to one of choosing between two composite hypotheses.

For each positive integer L, we denote by C^L the partitions of the label set consisting of L elements. In the decision space D, we assume that for each ω there is specified a nonempty subset D_ω of decisions that are viewed as "correct" when ω is true. We consider nonrandomized decision rules $\delta: \mathcal{X} \to D$, and define $C_\omega = \{x: \delta(x) \in D_\omega\}$. Any function Z on $\mathcal{X}$ may be termed a conditioning statistic. The range of Z is then the index B and $C^b = \{x: Z(x) = b\}$. A conditional confidence procedure is a pair (δ, Z). Its associated conditional confidence function (of b and ω) is the conditional density $\Gamma_\omega^b = P_\omega\{C_\omega | Z = b\}$. The classical symmetric C' rule is $\delta(X) = d_1$ when $X \geq 0$; i.e., $C_1 = \{x: x \geq 0\}$, and $P_\theta\{\delta(X) \in D_\theta\} = \Phi(|\theta|)$ for $|\theta| \geq \bar{\Delta}$. Often a value P* (close to one) is also specified, and the problem is phrased in terms of selecting the smallest n (in terms of P*, Δ^*) to yield $P_\theta\{\delta(X) \in D_\theta\} \geq P^*$ if $|\theta| \geq \bar{\Delta}$, which is satisfied if $P_{\bar{\Delta}}\{X > 0\} \geq P^*$.

Kiefer (1977) gives the following examples to illustrate some of the conditional confidence variations that are possible here.

6.2.2(a) Example. A procedure in C^2. With the same (C_0, C_1) as above, let $C^1 = \{x: |x| < r\}$ for some positive value r. Thus b = 1 and 2 fulfill the spirit of weaker and stronger conclusions as to which θ_i is larger. An easy computation using the monotone likelihood structure $(f_\theta(x)/f_\theta(-x)$ increases in $\theta > 0$ for each $x > 0)$ yields

(6.2.8)
$$|\theta| \geq \bar{\Delta} \Rightarrow$$
$$\Gamma_\theta^1 = [\Phi(r-|\theta|)-\Phi(-|\theta|)]/[\Phi(r-|\theta|)-\Phi(-r-|\theta|)] \geq \Gamma_{\bar{\Delta}}^1,$$
$$\Gamma_\theta^2 = [1+\Phi(-|\theta|-r)/\Phi(|\theta|-r)]^{-1} \geq \Gamma_{\bar{\Delta}}^2.$$

Hence the consumer can speak of conditional confidence "at least $\Gamma_{\bar{\Delta}}^b$ of selecting the larger mean when $|\theta| \geq \bar{\Delta}$," if $X \in C^b$. The values of n and r can be selected to yield satisfactory values of $\Gamma_{\bar{\Delta}}^1$ and $\Gamma_{\bar{\Delta}}^2$ and of the probabilities of correctly asserting these values.

6.2.2 (b) Example. Subset selection approach.

In this approach of Gupta (1956), the C' format allowed for three possible decisions (the nonempty subsets of {0,1}); the label of the larger population mean is asserted to be in the chosen subset. After reduction to (X,θ), this amounts to deciding whether θ is positive (d_1), negative (d_0), or real (d_2, say). The classical procedure modifies that of Example 6.2.2 (a) by making d_2 if $|X| < q$, for some positive q.

In the context of selection rules, some conditional procedures have been studied by Gupta and Huang (1975), Gupta, Huang and Huang (1976), Gupta and Nagel (1971) and Gupta and Wong (1976). All these papers deal with subset selection procedures for discrete distributions.

Gupta, Huang and Huang (1976) consider the conditional procedures for the k independent binomial populations $\pi_1,\ldots,\pi_k$ such that the independent observations X_i from π_i have frequency $b(n; p_i)$, $i = 1,\ldots,k$. A procedure of Gupta and Sobel type (1960), except that the rule is conditioned on the total number of observations $T = \sum_1^k X_i$, is studied. When k = 2, the infimum of the probability of correction selection (i.e. P(CS)) of this procedure is attained when $p_1 = p_2 = p$ and it is independent of p. The rule is defined as follows:

R_1: Select π_i if and only if
$$X_i \geq \max_{1 \leq j \leq k} X_j - d(t), \text{ given } \sum_1^k X_i = t.$$

Gupta, Huang and Huang (1976) derive a method leading to a conservative solution of the infimum of $P(CS|R_1)$ for the constant $d(t)$ depending on $T = t$ for the values of $k > 2$.

Gupta and Huang (1975) consider a selection problem for k independent Poisson populations with parameters $\lambda_1,\ldots,\lambda_k$. Suppose that we have equal sample size from each population. Let $X_1,\ldots,X_k$ denote the observed samples from the Poisson populations $\pi_1,\pi_2,\ldots,\pi_k$, respectively. Gupta and Huang (1975) propose the following rule R_2 to select the largest λ's:

R_2: Select π_i iff $X_i+1 \geq c(t) \max_{1\leq j\leq k} X_j$, given $\sum_1^k X_i = t$,

where $0 < c \leq 1$. This procedure R_2 can also be used to solve, conservatively, the problems of selection for a multinomial distribution. Gupta and Wong (1976) propose a procedure which modifies R_2 to select the smallest λ's.

An important purpose of conditional decision rules is to guarantee the same goal for unconditional rules as the conditional ones. Gupta and Nagel (1971) have proved the following result.

6.2.3 <u>Theorem</u>. Let $X_1, X_2, \ldots, X_k$ be independent and identically distributed random variables with joint distribution P_θ. Let $T(X_1, \ldots, X_k)$ be a sufficient statistic for θ.

(i) If $E(\delta(X_1, \ldots, X_k)|T) = P^*$ for all T then $E_\theta \delta = P^*$ for all θ.

(ii) If T is complete w.r.t. $\{P_\theta(\underline{x})\}$, then $E_\theta(\delta(X_1, \ldots, X_k)|T) = P^*$ is also necessary for $E_\theta \delta = P^*$ for all θ.

Gupta and Nagel (1971) have constructed tables necessary to carry out their conditional subset selection rules for the binomial, Poisson, negative binomial and Fisher's logarithmic distribution. For the binomial problem this randomized rule is as follows:

R_3:
$$\varphi_i(\underline{x}) = \begin{cases} 1 & \text{if } X_i > c_T \\ \rho & \text{if } X_i = c_T \\ 0 & \text{if } X_i < c_T \end{cases}$$

where $\rho = \rho(T, P^*, k)$ and $c_T = c_T(P^*, k)$ been computed and tabulated. Note that a procedure of this type will make the infimum of the probability of a correct selection equal to P^*.

6.3. <u>Multiple Comparison Procedures</u>

In 1966, Miller in his book <u>Simultaneous Statistical Inference</u> summarized what was then known about the theory and methods of multiple comparisons. Since that time research in this area has continued to yield a variety of new results. An overview and a bibliography of the developments during the past decade are given in Miller (1977). Although the literature on multiple hypothesis testing and multiple comparison methods is large, the literature on the optimality of the methods is rather scarce. An important contribution was made by Lehmann (1957). Lehmann (1957) finds optimal rules among the class of rules which are unbiased in a certain sense. Here optimality means minimizing the expected loss, and where the loss is the sum of the losses from the individual decisions.

It has been a common complaint that the powers of separate tests are small when using multiple tests. Therefore, in this section attention is directed towards maximizing the power of the individual test. Instead of using the constraint that the probability of at least one false rejection is smaller than a certain number α, an upper bound γ on the expected number of false rejections is used. Spjøtvoll (1972) shows that the latter is technically easier to work with and it is more instructive to think in terms of the expected number of false rejections than in terms of the probability of at least one false rejection. Suppose a statistician uses $\gamma = 0.05$, then on the average for every twentieth problem he makes one false statement. On the other hand if he uses $\alpha = 0.05$, then in the average for every twentieth problem he makes false rejections, but he does not know how many false rejections he makes. It is important to know this. It is also easily seen that the probability of at least one false rejection is less than γ, hence one has an upper bound on the probability of at least one false rejection when γ is known. The knowledge of α, however, cannot be used to give an upper bound on γ.

Let X be an observable random variable with probability distribution depending upon a parameter θ, $\theta \in \Omega$. Consider a family of hypothesis testing problems as follows:

(6.3.1) $\qquad H_t: \theta \in \Omega_{0t}$ vs $K_t: \theta \in \Omega_{1t}$, $t \in T$,

where $\Omega_{it} \subset \Omega$, $i = 0,1$ and T is finite with N elements. A test of the hypotheses (6.3.1) will be defined to be a vector $(\varphi_1(x), \ldots, \varphi_N(x))$, where the elements of the vector are ordinary test functions; when x is observed we reject H_t with probability $\varphi_t(x)$, $t \in T$. The power function of a test $(\varphi_1, \ldots, \varphi_N)$ is defined to be the vector $(\beta_1(\theta), \ldots, \beta_N(\theta))$ where $\beta_t(\theta) = E_\theta \varphi_t(X)$, $t \in T$. A related definition is the power of tests of a multiple hypothesis testing problem (see Duncan (1955)). Let $S(\gamma)$ be the set of all tests $(\varphi_1, \ldots, \varphi_N)$ such that

(6.3.2) $\qquad \sum_{t=1}^{N} E_\theta \varphi_t(X) \leq \gamma, \quad \theta \in \Omega_0,$

where $\Omega_0 = \cap_T \Omega_{0t}$. Hence $S(\gamma)$ is the set of tests such that the expected number of false rejections under Ω_0 is less than or equal to γ.

For each $t \in T$ we would, subject to (6.3.2), like to have $\beta_t(\theta)$ large when $\theta \in \Omega_{1t}$. If we make $\beta_t(\theta)$, $\theta \in \Omega_{1t}$, large for a fixed t, then $\beta_t(\theta)$, $\theta \in \Omega_{1t}$, will often have to be small for other values of t, if (6.3.2) is to be satisfied. Therefore, we will have to compromise, and we will consider tests which maximize the minimum power over certain subsets ω_t of Ω_{1t}, $t \in T$, and tests which maximize average power over certain subsets. A test $(\varphi_1, \ldots, \varphi_N) \in S(\gamma)$ will be said to maximize the minimum power over ω_t, $t \in T$, if it maximizes

(6.3.3) $\qquad \inf_T \inf_{\omega_t} E_\theta \psi_t(X)$

among tests $(\psi_1, \ldots, \psi_N) \in S(\gamma)$. It will be said to maximize the minimum average power over ω_t, $t \in T$, if it maximizes

(6.3.4) $$\sum_{t=1}^{N} \inf_{\omega_t} E_\theta \psi_t(X)$$

among tests $(\psi_1,\ldots,\psi_N) \in S(\gamma)$.

Note that the above optimality criteria are more directed towards the performances of the individual tests, than towards their simultaneous performance. Let, for example, X_1 and X_2 be independent normal $N(\mu_i,1)$, $i = 1,2$ and $H_i: \mu_i = 0$ vs $K_i: \mu_i > 0$, $i = 1,2$. Furthermore, let $\omega_i = \{(\mu_1,\mu_2)|\mu_i \geq \Delta\}$, $i = 1,2$, for some $\Delta > 0$. Then a test of the two hypotheses satisfying (6.3.3) is such that if one of the μ_i is greater than Δ, then we have a guaranteed minimum probability of discovering this, and this minimum probability is the largest possible. The optimality criterion does not tell us anything about the probability of rejecting both H_1 and H_2 when both μ_1 and μ_2 are greater than Δ. The second criterion (6.3.4) is similarly directed towards maximizing the minimum average of the powers of the individual tests. The reason for studying individual powers is that a common complaint about multiple comparison methods has been that the individual powers are very small. It is the objective of the procedures of Spjøtvol (1972) to maximize the power of individual tests. Several examples have been given by Spjøtvol (1972) to show the application of the proposed procedures.

Let $f_{01},\ldots,f_{0N},f_1,\ldots,f_N$ be (known) integrable functions with respect to a σ-finite measure μ defined on a measurable space $(\mathcal{X},\hat{\mathcal{U}})$. Let $S'(\gamma)$ be the set of all tests $(\psi_1,\ldots,\psi_N)$ satisfying

$$\sum_{t=1}^{N} \int \psi_t(x) f_{0t}(x) d\mu(x) = \gamma.$$

Suppose that there exists a test $(\varphi_1,\ldots,\varphi_N) \in S'(\gamma)$ defined by

(6.3.5) $$\varphi_t(x) = \begin{cases} 1 & \text{when } f_t(x) > cf_{0t}(x), \\ a_t & \text{when } f_t(x) = cf_{0t}(x), \\ 0 & \text{when } f_t(x) < cf_{0t}(x). \end{cases}$$

Then it has been shown by Spjøtvol (1972) that $(\varphi_1,\ldots,\varphi_N)$ maximizes

(6.3.6) $$\sum_{t=1}^{N} \int \psi_t(x) f_t(x) d\mu(x)$$

among all tests $(\psi_1,\ldots,\psi_N) \in S'(\gamma)$. He has given the following examples to show its applications.

Application to comparison of means of normal random variables with common known variance.

Let X_{ij} be normal $N(\mu_i,1)$, $j = 1,\ldots,n_i$, $i = 1,\ldots,r$, and independent. Hence we assume, without loss of generality, that the known variance is 1. Consider the following hypotheses about linear functions in the $\mu_1,\ldots,\mu_r$,

(6.3.7) $$H_t: \sum_{i=1}^{r} a_{ti}\mu_i = b_t \text{ vs } K_t: \sum_{i=1}^{r} a_{ti}\mu_i > b_t,$$

$t \in T$, where the a_i and b_i, $1 \leq i \leq r$, are given constants. Let ω_t be the set of all parameter points such that

$$\sum_{i=1}^{r} a_{ti}\mu_i - b_t \geq \Delta_t,$$

where $\Delta_t > 0$ is to be fixed later. We assume that Ω_0 is not empty.

We know that the condition (6.3.2) on the tests $\{\varphi_t\}$ becomes

(6.3.8) $\qquad \sum_{t=1}^{N} \Phi[-\frac{1}{2}\Delta_t(\sum_{i=1}^{r} a_{ti}^2/n_i)^{-\frac{1}{2}} - \Delta_t^{-1}k_t(\sum_{i=1}^{r} a_{ti}^2/n_i)^{\frac{1}{2}}] \leq \gamma,$

where

$$\varphi_t(x) = \begin{cases} 1 & \text{if } (\sum_{i=1}^{r} a_{ti}^2/n_i)^{-\frac{1}{2}}(\sum_{i=1}^{r} a_{ti}\bar{X}_i - b_t) \\ & \geq \frac{1}{2}(\sum_{i=1}^{r} a_{ti}^2/n_i)^{-\frac{1}{2}}\Delta_t + \Delta_t^{-1}k_t(\sum_{i=1}^{r} a_{ti}^2/n_i)^{\frac{1}{2}}, \\ 0 & \text{otherwise,} \end{cases}$$

and $\bar{X}_i = \frac{1}{n_i}\sum_{j=1}^{n_i} X_{ij}$, $i = 1,\ldots,r$, $k_t = -\log c_t$, $t = 1,\ldots,T$. It is seen that the test maximizing the minimum average power over the alternatives ω_t is given by the tests $\{\varphi_t\}$ with $k_t = k$, $t \in T$, and where k is determined so that we have equality in (6.3.8).

Spjøtvol (1972) discusses various special cases as follows:

(1) <u>Differences between means</u>. Here the hypotheses are

(6.3.9) $\qquad H_{ij}: \mu_i = \mu_j$ vs $K_{ij}: \mu_i > \mu_j$, $i \neq j$.

The pair (i,j) corresponds to the index t, and $N = r(r-1)$. Note that H_{ij} vs K_{ij} and H_{ji} vs K_{ji} are two different problems; H_{ij} is the same as H_{ji}, but the alternatives are different. If we have used alternatives $\mu_i \neq \mu_j$, T would have $\frac{1}{2}r(r-1)$ elements. But since one usually wants to know which mean is the greater when H_{ij} is rejected, the above formulation of the problem seems to be the more useful one. In general we find that the test maximizing minimum average power and minimum power over the alternatives

$$\mu_i - \mu_j \geq \Delta(\frac{1}{n_i} + \frac{1}{n_j})^{\frac{1}{2}},$$

rejects H_{ij} and accepts K_{ij} when

$$(\frac{1}{n_i} + \frac{1}{n_j})^{-\frac{1}{2}}(\bar{X}_i - \bar{X}_j) > Z\rho,$$

where $\rho = \gamma/r(r-1)$.

(2) <u>Comparison with a known standard</u>.

We want to compare the means $\mu_1,\ldots,\mu_r$ with a known standard μ_0. More precisely our problem is

$$H_t: \mu_t = \mu_0 \text{ vs } K_t: \mu_t > \mu_0, \quad t = 1,\ldots,r.$$

In this case $N = r$, and the test maximizing minimum average power and minimum power over alternatives of the form $\mu_t - \mu_0 \geq \Delta n_t^{\frac{1}{2}}$, $t = 1,\ldots,r$, consists of rejecting H_t when $n_t^{\frac{1}{2}}(\bar{X}_t - \mu_0) > z_\rho$, where $\rho = \frac{\gamma}{r}$. We could, of course, also have added hypotheses with alternatives with $\mu_t < \mu_0$ if that was a possible result, and was of interest to the experimenter. In that case $N = 2r$ and $\rho = \gamma/2r$.

(3) <u>Comparison with an unknown standard or a control</u>.

In this case one of the means, μ_r, is a control. The problem is

$$H_t: \mu_t - \mu_r = 0 \text{ vs } K_t: \mu_t - \mu_r > 0,$$

$t = 1,\ldots,r-1$. If we consider alternatives of the form $\mu_t - \mu_r \geq \Delta(\frac{1}{n_t} + \frac{1}{n_r})^{\frac{1}{2}}$, we will reject H_t when $(\frac{1}{n_t} + \frac{1}{n_r})^{-\frac{1}{2}}(\bar{X}_t - \bar{X}_r) > z_\rho$, where $\rho = \gamma/(r-1)$. Again we could have considered alternatives $\mu_t - \mu_r < 0$ at the expense of decreasing ρ.

4. <u>Ordered means</u>. In some situations it is known that $\mu_1 \leq \mu_2 \leq \ldots \leq \mu_r$. Then we would consider the problem (6.3.9) with $i > j$. We get the same results with $\rho = 2\gamma/r(r-1)$ instead of $\rho = \gamma/r(r-1)$. Hence the power of the test is larger in this case.

It should be pointed out that the multiple comparison procedures sometimes have difficulties to derive conclusions after pairwise comparisons. The deficiencies are also inherited from the framework of NP formulation. Gupta and Huang (1977) have used some optimal criteria to derive optimal selection procedures based on tests. This work partially remedies the deficiencies above.

Gupta and Huang (1977) have studied, as an example, for k independent normal populations $\pi_1,\ldots,\pi_k$. Let $X_{i1},\ldots,X_{in}$ be independent random observations with normal distributions $N(\theta_i,1)$, $i = 1,\ldots,k$. Let $\underline{\theta} = (\theta_1,\ldots,\theta_k)$,

$$\Omega_i = \{\underline{\theta} | \theta_i \geq \max_{j \neq i} \theta_j + \Delta\}, \quad i = 1,\ldots,k,$$

and

$$\Omega_0 = \{\underline{\theta} | \theta_1 = \ldots = \theta_k\}$$

where Δ is a given positive constant.

Now, the problem is to consider a family of hypotheses testing as follows:

(6.3.10) $\qquad H_{0i}: \underline{\theta} \in \Omega_0 \text{ vs } H_i: \underline{\theta} \in \Omega_i,$

$i = 1,\ldots,k$. A test of the hypotheses (6.3.10) will be defined to be a vector $(\delta_1(\underline{x}),\ldots,\delta_k(\underline{x}))$, where $\underline{X} = (X_1,\ldots,X_k)$, $X_i = \frac{1}{n}\sum_{j=1}^{n} X_{ij}$, $1 \leq i \leq k$, and $\underline{X} = \underline{x}$, the elements of the vector are ordinary test functions; when $\underline{x}$ is observed we reject H_{0i} with probability $\delta_i(\underline{x})$, $1 \leq i \leq k$. The power function of a test $(\delta_1,\ldots,\delta_k)$ is defined to be the vector $(\beta_1(\underline{\theta}),\ldots,\beta_k(\underline{\theta}))$, where $\beta_i(\underline{\theta}) = E_{\underline{\theta}}\delta_i(\underline{X})$, $1 \leq i \leq k$. In other words, for $\underline{\theta} \in \Omega_i$, we know that $\beta_i(\underline{\theta})$ is the probability of correct selection and $\delta_i(\underline{x})$ is the probability of selecting the best population π_i. Let S_γ be the set of all tests $(\delta_1,\ldots,\delta_k)$ such that

(6.3.11) $$\sum_{i=1}^{k} E_{\underline{\theta}}\delta_i(\underline{X}) \leq \gamma, \quad \underline{\theta} \in \Omega_0.$$

We explain the expected subset size for the selection rule $\delta = (\delta_1,\ldots,\delta_k)$ over Ω_0 in S_γ as the error of false rejection is less than or equal to γ.

Let $S'_\gamma = \{\delta = (\delta_1,\ldots,\delta_k) | \sum_{i=1}^{k} E_{\underline{\theta}_0}\delta_i(\underline{x}) = \gamma\}$,

where $\underline{\theta}_0 = (\theta_0,\ldots,\theta_0)$. Gupta and Huang (1977) have proved that let $\delta^0 = (\delta_1^0,\ldots,\delta_k^0) \in S'_\gamma$ be defined by

$$\delta_i^0(\underline{x}) = \begin{cases} 1 & \text{if } x_i \geq \frac{1}{k-1}\sum_{j \neq i} x_j + c, \\ 0 & \text{if } x_i < \frac{1}{k-1}\sum_{j \neq i} x_j + c, \end{cases}$$

then δ^0 maximizes $\sum_{i=1}^{k} \inf_{\underline{\theta} \in \Omega_i} E_{\underline{\theta}}\delta_i(\underline{X})$ among all rules $\delta \in S_\gamma$.

Note that we have compared any population π_i with all remaining π_j's at a time. There is not any difficulty for the comparable problems as in the case of multiple comparison rules. We can obtain the conclusions after we apply the selection rules. The selection approach for the problems of multiple comparisons is a very efficient way to do them.

6.4 Multiple Range Tests

Let $X_1,\ldots,X_s$ be independently distributed with distribution

(6.4.1) $$P(X_i \leq x) = F(x-\theta_i),$$

where F is known and continuous, and let Y_i denote the ith smallest of the X's. Lehmann and Shaffer (1977) discuss the problem of grouping the θ's by means of a multiple-range test. The following definition of such a procedure in terms of critical values $C_2,\ldots,C_s$ is adopted by many authors.

As a first step the range $R_s = Y_s - Y_1$ is compared with C_s: If $R_s < C_s$, the θ's are declared indistinguishable and the procedure terminates; if $R_s \geq C_s$, the θ's corresponding to Y_1 and Y_s are declared to differ, and the two $(s-1)$ ranges,

$$R_{s-1,1} = Y_{s-1} - Y_1 \text{ and } R_{s,2} = Y_s - Y_2,$$

are compared with C_{s-1}. If both R's are less than C_{s-1}, the two sets of θ's corresponding to $(Y_1, Y_2,\ldots,Y_{s-1})$ and $(Y_2,\ldots,Y_s)$ are declared indistinguishable and the procedure terminates. Otherwise, the two θ's corresponding to Y_1, Y_{s-1} or the two θ's corresponding to Y_2, Y_s are declared to differ, and the three or appropriate (s-2) ranges are compared with C_{s-2} and so on. After a set of means has been declared indistinguishable, all of its subsets are also considered indistinguishable without further test.

To complete specification of the procedure, it is necessary to decide on the critical values $c_2,\ldots,c_s$. This choice is typically made in terms of the probabilities

(6.4.2) $$\alpha_k = P(R_k \geq C_k),$$

where R_k denotes the range of k independent X's with common location parameter θ. Two well-known choices of the α's are those of Newman-Keuls (1939, 1952) given by

(6.4.3) $$\alpha_2 = \ldots = \alpha_s$$

and of Duncan (1955) given by

(6.4.4) $$\alpha_i = 1-(1-\alpha)^{i-1}.$$

The definition of a multiple-range test given in the preceding is that adopted by, among others, Dunnett (1970), Hartley (1955), Keuls (1952), O'Neil and Wetherill (1971), Ryan (1959) and Spjøtvol (1974). However, some other authors, among them Tukey (1953), Miller (1966) and Einot and Gabriel (1975), add to this definition the requirement that a set of means is declared inhomogeneous only if all sets of means containing S are also declared inhomogeneous. Imposing this restriction on a procedure with arbitrary critical values $C_2,\ldots,C_s$ is equivalent to replacing these values by

(6.4.5) $$C'_j = \max(C_2,\ldots,C_j), \quad j = 2,\ldots,s.$$

These new values will agree with the original ones if and only if the latter satisfy the monotonicity condition.

(6.4.6) $$C_2 \leq \ldots \leq C_s.$$

When (6.4.6) does not hold, replacement of the C_i by the C'_i, of course, affects the probabilities α_i (and imposition of the restriction (6.4.6) limits the rates at which the α's can increase). Lehmann and Shaffer (1977) point out that a procedure which is caught in this dilemma is that of Duncan (1955), who defines his α's by (6.4.4) but in his tables imposes (6.4.6) whenever this condition is violated. He is thus actually considering two distinct procedures without ever quite reconciling this difference. However, the Newman-Keuls procedure does not run into this difficulty since the C's in that case obviously satisfy (6.4.6).

Lehmann and Shaffer (1977) have established two properties of multiple-range tests, which can be stated generally in the following lemma.

6.4.1 <u>Lemma</u>. Let R'_k denote the range of $(X_1,\ldots,X_k)$, $(k \leq s)$. Then: (i) If

(6.4.7) $$C_k \leq C_\ell \quad \text{for all } \ell > k,$$

the set $\{\theta_1,\ldots,\theta_k\}$ can only be declared inhomogeneous when $R'_k \geq C_k$.
(ii) If

(6.4.8) $C_\ell \leq C_k$ for all $\ell > k$,

the set $\{\theta_1,\ldots,\theta_k\}$ will be declared inhomogeneous whenever $R'_k \geq C_k$.

Proof. (i) Suppose that $R'_k < C_k$ and let $k' > k$ be the number of X_j's spanned by $X_1,\ldots,X_k$, i.e. satisfying

(6.4.9) $\min(X_1,\ldots,X_k) \leq X_j \leq \max(X_1,\ldots,X_k)$.

Then at the stage of the procedure in which the ranges of k' variables are considered, we find $\min(X_1,\ldots,X_k)$ and $\max(X_1,\ldots,X_k)$ as the end point of such a range with

$$R'_k = \max(X_1,\ldots,X_k) - \min(X_1,\ldots,X_k) < C_k \leq C_{k'}.$$

The set $\{\theta_1,\ldots,\theta_k\}$ will, therefore, be declared to be indistinguishable if it has not already been declared so at an earlier stage.

Note that (6.4.7) is needed in this argument, since when (6.4.7) does not hold, rejection can occur even when $R'_k < C_k$. Suppose, e.g., that $s = 3$, $C_2 > C_3$, and $k = 2$; and consider the case that

$$X_1 < X_3 < X_2 \text{ and } C_3 < X_2 - X_1 < C_2.$$

Then at the first step of the procedure, $X_2 - X_1$ is compared with C_3, and hence θ_1 and θ_2 are declared to differ in spite of the fact that $|X_2 - X_1| < C_2$.

(ii) Suppose first that $k' = s$, so that (6.4.9) holds for all $j = 1,\ldots,s$. Then the θ's corresponding to $\min(X_1,\ldots,X_k)$ and $\max(X_1,\ldots,X_k)$ will be declared to differ at the first step since $R'_k \geq C_k \geq C_s$.

Suppose next that $k' = s-1$. Then the set $\{\theta_1,\ldots,\theta_s\}$ will be declared inhomogeneous at the first step since its range exceeds $R'_k \geq C_k \geq C_s$, and then $\{\theta_1,\ldots,\theta_k\}$ will be declared inhomogeneous at the second step. Continuation of this argument establishes the lemma.

Of central interest to many investigators of multiple-range tests is the probability of declaring at least one pair of θ's to differ when the members of that pair are in fact equal, say

(6.4.10) $\alpha(\theta_1,\ldots,\theta_s) = P_{\theta_1,\ldots,\theta_s}$ [at least one false significance statement].

Suppose that the θ's fall into t groups of equal values of lengths $v_1,\ldots,v_t$. Without loss of generality, let the θ's be numbered so that

(6.4.11) $\theta_1 = \ldots = \theta_{v_1}; \theta_{v_1+1} = \ldots = \theta_{v_1+v_2}; \ldots;$

with the values in the different groups being distinct.

For this situation the following theorem of Lehmann and Shaffer (1977) applies:

6.4.2 Theorem. The inequalities (6.4.6) constitute a necessary and sufficient condition for

(6.4.12) $\sup \alpha(\theta_1,\ldots,\theta_s) = 1 - \prod_{i=1}^{t}(1-\alpha_{v_i})$, (the sup is taken over all θ's which after reordering satisfy (6.4.11) and where $\alpha_1 = 0$.) to hold for all configurations $(v_1,\ldots,v_t)$ and all values (6.4.11).

Proof. (a) Sufficiency. Let R_i' denote the range of the X's corresponding to the ith group of θ's, and suppose that (6.4.6) holds. Then (6.4.7) also holds, and it follows from part (i) of Lemma 6.4.1 that false rejection can occur only when at least one of the ranges R_i' satisfies $R_i' \geq C_{v_i}$. Thus

(6.4.13) $$\alpha(\theta_1,\ldots,\theta_s) \leq P(\bigcup_{i=1}^{t} [R_i' \geq C_{v_i}]).$$

The right side of (6.4.13) is equal to

(6.4.14) $$1 - P(R_i' \leq C_{v_i} \text{ for all } i).$$

This equals the right side of (6.4.12) which is, therefore, an upper bound for $\alpha(\theta_1,\ldots,\theta_s)$.

To prove that it is a sharp upper bound, suppose that the difference between the values of successive groups in (6.4.11) is Δ. Then as $\Delta \to \infty$, the probability $\alpha(\theta_1,\ldots,\theta_s)$ tends to the right side of (6.4.12) and this completes the proof.

(b) Necessity. Let k be the smallest integer for which $C_{k+1} < C_k$, so that

$$C_2 \leq \ldots \leq C_{k-1} \leq C_k \text{ and } C_{k+1} < C_k.$$

Consider a configuration of θ's for which

(6.4.15) $$\theta_1 = \ldots = \theta_k < \theta_{k+1},$$

and the remaining θ's are all different and tend to infinity. Then all ranges involving one of the variables $(X_1,\ldots,X_{k+1})$ and one of the variables $(X_{k+2},\ldots,X_s)$ will exceed $C_k,\ldots,C_s$ and thus will not affect the error probabilities. We shall, therefore, without loss of generality suppose that $s = k+1$.

For the configuration (6.4.15), the probability of at least one false significance statement is the probability of declaring at least one pair from $\theta_1,\ldots,\theta_k$ significant. We shall now show that this probability exceeds α_k. The proof of this result will consist of two steps:

(i) It follows from part (ii) of the Lemma 6.4.1 that at least one pair from $\theta_1,\ldots,\theta_k$ will be declared significant when the range $R(X_1,\ldots,X_k)$ is $\geq C_k$. (ii) In addition we shall show that there are cases with $R(X_1,\ldots,X_k) < C_k$ for which at least one pair from $(\theta_1,\ldots,\theta_k)$ will be declared significant. Since the probability of the first case is α_k, this will show that the probability of at least one pair from $(\theta_1,\ldots,\theta_k)$ being declared significant exceeds α_k, as was to be proved.

To see that the possibilities (i) do not exhaust the cases in which $(\theta_1,\ldots,\theta_k)$ is declared inhomogeneous, consider the cases in which

$$C_{k+1} < R(X_1,\ldots,X_k) < C_k$$

and

$$\min(X_1,\ldots,X_k) < X_{k+1} < \max(X_1,\ldots,X_k).$$

Then, as before, the θ's corresponding to $\min(X_1,\ldots,X_k)$ and $\max(X_1,\ldots,X_k)$ will be declared to differ. This completes the proof.

Lehmann and Shaffer (1977) propose a result which is closely related to Theorem 6.3.2 and also mentioned by both Tukey (1953) and Einot and Gabriel (1975) and which provides an added interpretation of the probabilities α_k.

6.4.3 <u>Theorem</u>. Consider all $(\theta_1,\ldots,\theta_s)$ for which

(6.4.16) $$\theta_1 = \ldots = \theta_k.$$

Condition (6.3.6) is necessary and sufficient for

(6.4.17) $\quad \alpha_k = \sup P[\text{falsely rejecting at least one of the equalities}$
$\quad\quad\quad\quad (6.4.16)]$

to hold for all k. Here the sup is taken over all $(\theta_1,\ldots,\theta_s)$ satisfying (6.4.16).

<u>Proof</u>. <u>Sufficiency</u>. Suppose that (6.4.6) holds. That α_k is an upper bound to the probabilities on the right side of (6.4.17) follows from part (i) of Lemma 6.4.1. That the upper bound is sharp is seen from Theorem 6.4.2 by letting the distances of the remaining θ's from each other and from the value (6.4.16) all tend to infinity.

<u>Necessity</u>. The proof is the same as that of the necessity part of Theorem 6.4.2.

6.5 <u>Multistage Comparison Procedures</u>

Consider s independent random quantities X_i, $i = 1,\ldots,s$ (they may, for example, be vector-valued) with distributions P_i ranging over sets $\mathcal{P}_i$. Lehmann and Schaffer (1979) consider the problem of distinguishing among functions g_i defined over $\mathcal{P}_i$ by means of a multistage procedure based on the X_i and define the following rule.

A multistage procedure is defined through a series of rejection regions. The first stage consists of a test of the hypothesis

(6.5.1) $$H_s: \quad g_1(P_1) = \ldots = g_s(P_s)$$

by means of a rejection region R_s in the sample space of $(X_1,\ldots,X_s)$. If $(X_1,\ldots,X_s) \in R_s$, the hypothesis H_s is rejected, or, as we shall say, the homogeneity of the set $\{g_1,\ldots,g_s\}$ is rejected and the procedure goes on to stage 2 in order to get more detail about the reasons for rejection. If $(X_1,\ldots,X_s)$ falls into the complement $\bar{R}_s$ of R_s, we shall employ the traditional terminology of hypothesis testing and say that H_s or the homogeneity of the set $\{g_1,\ldots,g_s\}$ is accepted. This, of course, is not meant to imply that the data have convinced us of the validity of (6.5.1) but only that they have not enabled us to reject (6.5.1). In that case no further tests will be made.

The second stage consists of tests of the s hypotheses

(6.5.2) $$H_{s-1;i}: \quad g_1 = \ldots = g_{i-1} = g_{i+1} = \ldots = g_s$$

by means of rejection regions $R_{s-1;i}$. If the (s-1)-tuple $(X_1,\ldots,X_{i-1},X_{i+1},\ldots,X_s)$ falls into the complement of $R_{s-1;i}$, homogeneity of the set $\{g_1,\ldots,g_{i-1},g_{i+1},\ldots,g_s\}$ is accepted and no subset is tested. If on the other hand $(X_1,\ldots,X_{i-1}, X_{i+1},\ldots,X_s)$

falls into $R_{s-1;i}$, homogeneity of the set $\{g_1,\ldots,g_{i-1},g_{i+1},\ldots,g_s\}$ is rejected and one proceeds to the third stage.

This procedure is continued until nothing is left to be tested. As a shorthand notation we shall sometimes write H_k and R_k instead of the more complete $H_{k;i_1,i_2,\ldots}$ and $R_{k;i_1,i_2,\ldots}$ when the additional subscripts are clear from the context.

As an illustration, Lehmann and Schaffer (1979) give in Figures 6.5(a) and 6.5(b) examples of possible outcomes of such a procedure for the case s = 4 where homogeneity has been accepted for the underlined sets and rejected for those not underlined.

$$\begin{array}{cccccc} & & 1\ 2\ 3\ 4 & & & \\ & 123 & 124 & 134 & 234 & \\ \underline{12} & 13 & 14 & \underline{23} & 24 & 34 \end{array}$$

Fig. 6.5(a)

$$\begin{array}{cccccc} & & 1\ 2\ 3\ 4 & & & \\ & \underline{123} & \underline{124} & 134 & \underline{234} & \\ \underline{12} & \underline{13} & \underline{14} & \underline{23} & 24 & \underline{34} \end{array}$$

Fig. 6.5(b)

Typically, the rejection regions are defined in terms of statistics T such as ranges or χ^2-statistics in the case of normal means, Kruska-Wallis and Smirnov statistics in the nonparametric case, etc. These are computed for the variables in question and provide measures of inhomogeneity. Then R_s is the set for which $T_s(X_1,\ldots,X_s) \geq C_s$; $R_{s-1;i}$ is the set for which $T_{s-1;i}(X_1,\ldots,X_{i-1},X_{i+1},\ldots,X_s) \geq C_{s-1;i}$, and so on. For the T's and C's we shall frequently delete the subscripts behind the semi-colon as we did for the H's and R's. In the most commonly used procedures, the statistics T are obtained from a single functional (for example, the range) evaluated at the respective empirical cumulative distribution functions, but this need not be the case and will not be assumed here.

It follows from the definition of stagewise procedures given above that they have the following properties.

(i) Rejection of homogeneity for any set of g's implies rejection of homogeneity for all sets containing it.

(ii) Acceptance of homogeneity for any set of g's implies acceptance of homogeneity for any set contained in it.

(iii) Homogeneity of the set $\{g_{i_1},\ldots,g_{i_k}\}$ can be rejected only when $X_{i_1},\ldots,X_{i_k}$ falls into R_k.

6.5.1 <u>Definition</u> (Lehmann and Shaffer (1979)).

Suppose the g's fall into t distinct groups of sizes $v_1, v_2,\ldots,v_t$ respectively ($\Sigma v_i = s$), say,

(6.5.2) $$g_{i_1} = \ldots = g_{i_{v_1}} \; ; \; g_{i_{v_1+1}} = \ldots = g_{i_{v_1+v_2}} \; ; \ldots$$

where $(i_1, i_2, \ldots, i_s)$ is a permutation of $(1, \ldots, s)$. We shall say that the configuration (6.5.2) is separable by a given multistage procedure if there exists a sequence of distribution $(P_1^{(m)}, \ldots, P_s^{(m)})$ with $P_i^{(m)} \in P_i$ and satisfying (6.5.2) such that any hypothesis

(6.5.3) $$g_{j_1} = \ldots = g_{j_r}$$

with subscripts from at least two of the t distinct subsets of (6.5.2) is rejected with probability tending to 1 as $m \to \infty$. If every configuration (6.5.2) is separable by a given multistage procedure, we call the procedure separating.

Lehmann and Shaffer (1979) give an example to show the separability as follows.

6.5.2 Example. Let $X_i = (X_{i1}, \ldots, X_{in})$ with the X_{ij} independently normal distributed with mean θ_i and known common variance σ^2. Let $g_i(P_i) = \theta_i$ and let $X_i.$ denote the mean of the ith sample. If each of the statistics $T_s, T_{s-1}, \ldots$ is, for example, the range of the $X_i.$'s in question, or the sum of the squared deviations of the $X_i.$'s from their mean, the procedure is clearly separating since such a T tends in probability to infinity if the difference between at least two of the involved θ's tends to infinity.

The choice of the critical values C_k is typically made in terms of the probability of rejecting homogeneity for the subset $\{g_{i_1}, \ldots, g_{i_k}\}$ when in fact

(6.5.4) $$g_{i_1} = \ldots = g_{i_k}.$$

The value of this probability depends also on the remaining g's or P's since an earlier hypothesis may be accepted thereby depriving the hypothesis (6.5.4) of the chance of being tested. A quantity of interest in this connection is

(6.5.5) $$\alpha_k = P_{g_{i_1} = \ldots = g_{i_k}}[(X_{i_1}, \ldots, X_{i_k}) \in R_k],$$

where R_k may, of course, also depend on $i_1, \ldots, i_k$ but where we shall assume that the probability on the right hand side is the same for all $(P_{i_1}, \ldots, P_{i_k})$ satisfying (6.5.4).

It follows from property (iii) that

(6.5.6) $$P[\text{rejecting homogeneity of } \{g_{i_1}, \ldots, g_{i_k}\}] \leq \alpha_k$$

for all $\{g_1, \ldots, g_s\}$ satisfying (6.5.4). Consider now the configuration (6.5.2) with $v_1 = k$ and $v_2 = v_3 = \ldots = 1$, and suppose this configuration is separable by the given procedure. Then there exists a sequence of distributions $(P_1^{(m)}, \ldots, P_s^{(m)})$ satisfying (6.5.2) and for which the left hand side of (6.5.6) will tend to α_k. It follows that

(6.5.7) $$\sup P[\text{rejecting homogeneity of } \{g_{i_1}, \ldots, g_{i_k}\}] = \alpha_k$$

where the sup is taken over all $(P_1, \ldots, P_s)$ satisfying (6.5.2). The relation (6.5.7) thus holds in particular whenever the given procedure is separating. For the normal case this result was already noted by Tukey in Chapter 30 (1953). The levels α_k are called apparent significance levels by Tukey (1953), k-mean significance levels by Duncan (1955), and nominal levels by Einot and Gabriel (1975).

An aspect of a multiple comparison procedure which is of great interest is the maximum probability of rejecting the homogeneity of at least one set of g's which are in fact equal, say

(6.5.8) $$\alpha_0 = \sup \alpha(P_1,\ldots,P_s)$$

where for a given s-tuple $(P_1,\ldots,P_s)$

(6.5.9) $$\alpha(P_1,\ldots,P_s) = P[\text{at least one false rejection}]$$

and where the sup in (6.5.8) is taken over all possible distributions $P_i \in \mathcal{P}_i$.

Lehmann and Shaffer (1979) have proved the following result to show the relation of α_0 to the significance levels $\alpha_2,\ldots,\alpha_s$.

6.5.3 <u>Theorem</u>. If the configuration of $(g_1,\ldots,g_s)$ is given by (6.5.1) and is separable by the given procedure, then

(6.5.10) $$\sup \alpha(P_1,\ldots,P_s) = 1 - \prod_{i=1}^{t} (1-\alpha_{v_i})$$

where $\alpha_1 = 0$ and where the supremum is taken over all s-tuples $(g_1,\ldots,g_s)$ which satisfy (6.5.1).

The proof of this theorem is essentially the same as that given for Theorem 6.5.2 and will, therefore, be omitted here. It follows from (6.5.10) that

(6.5.11) $$\alpha_0 = \sup_{v_1,\ldots,v_t} \cdot [1- \prod_{i=1}^{t} (1-\alpha_{v_i})]$$

holds for all configurations provided the given procedure is separating. It may be of interest to note that separability of the configuration is not only sufficient for Theorem 6.5.3 but also necessary.

The remarkable feature of these results is the fact that the suprema in question are independent of the statistics T defining the procedure, and depend only on the values $\alpha_2,\ldots,\alpha_s$. Our primary concern is the choice of these α's.

In analogy with standard practice in hypothesis testing, we shall impose a bound on one of the rejection probabilities, namely α_0, thereby insuring comparability of different procedures. We shall fix a value, say α_0^* and restrict attention to procedures satisfying

(6.5.12) $$\alpha_0 \leq \alpha_0^*.$$

Subject to this condition, we should like to maximize the "power" of the procedure, that is, the probability of detecting existing inhomogeneities. Since we are assuming the statistics T to be given and only the critical values C to be at our disposal, it is clear that the power is maximized by maximizing the α's.

From (6.5.11) we see that the problem is that of maximizing $\alpha_2,\ldots,\alpha_s$ subject to

(6.5.13) $$\prod_{i=1}^{t}(1-\alpha_{v_i}) \geq 1-\alpha_0^* \quad \text{for all } v_1,\ldots,v_t \text{ satisfying } \sum_{i=1}^{t} v_i = s.$$

Let us begin with α_s. If the left hand side of (6.5.13) is to involve α_s, we must have $t = 1$, $v_1 = s$ and the only restriction on α_s is $\alpha_s \leq \alpha_0^*$, so that we shall maximize α_s by putting

(6.5.14) $$\alpha_s = \alpha_0^*.$$

Analogously, we see that we can also put

(6.5.15) $$\alpha_{s-1} = \alpha_0^*$$

without imposing any additional restriction on $\alpha_2,\ldots,\alpha_{s-2}$. This shows that for s=3, there exists a uniformly best choice of the α's, namely $\alpha_2 = \alpha_3 = \alpha_0^*$.

Consider next the case $s = 4$. The possible configurations are $t = 1$, $v_1 = 4$; $t = 2$, $\{v_1,v_2\} = \{1,3\}$ or $\{2,2\}$; $t = 3$, $\{v_1,v_2,v_3\} = \{1,1,2\}$ and $t = 4$, $v_1 = v_2 = v_3 = v_4 = 1$ and it is seen that a uniformly best choice of the α's is given by $\alpha_3 = \alpha_4 = \alpha_0^*$, $\alpha_2 = 1-(1-\alpha_0^*)^{\frac{1}{2}}$.

This happy state of affairs, however, does not extend to higher values of s. For $s = 5$, for example, it is no longer possible simultaneously to maximize α_2 and α_3. The maximum value of α_3 is α_0^* but this requires $\alpha_2 = 0$; alternatively, the maximum value of α_2 is $1-(1-\alpha_0^*)^{\frac{1}{2}}$ in which case α_3 cannot exceed $1-(1-\alpha_0^*)^{\frac{1}{2}}$.

When a uniformly best choice of the α's does not exist, for convenience, Lehmann and Shaffer (1979) introduce the following definition.

6.5.4 Definition. A set $(\alpha_2,\ldots,\alpha_s)$ for which α_0 satisfies (6.5.12) will be called inadmissible if there exists another set $(\alpha_2',\ldots,\alpha_s')$ for which α_0' also satisfies (6.5.12) and such that

(6.5.16) $\alpha_i \leq \alpha_i'$ for all i, with strict inequality for at least some i.

This use of the term is analogous to that which is customary in the theory of hypotheses-testing.

Lehmann and Shaffer (1979) propose the following two theorems to provide some guidance for the choice of the levels $\alpha_2,\ldots,\alpha_s$ and α_0.

6.5.5 Theorem. Any admissible procedure which is separating satisfies

(6.5.17) $$\alpha_2 \leq \alpha_3 \leq \cdots \leq \alpha_s.$$

Proof. If (6.5.17) does not hold, there exists a k such that $\alpha_{k+1} < \alpha_k$. For any such k, consider the procedure in which $\alpha_i' = \alpha_i$ for $i \neq k+1$ and $\alpha_{k+1}' = \alpha_k$. Then clearly $\alpha_0' \geq \alpha_0$. To show that $\alpha_0' \leq \alpha_0^*$, we need only show that $\prod_{i=1}^{t} (1-\alpha_{v_i}') \geq 1-\alpha_0^*$ for all $(v_1,\ldots,v_n)$.

If none of the v's is equal to k+1, $\alpha_{v_i}' = \alpha_{v_i}$ for all $i = 1,\ldots,t$ and the result holds. Otherwise, replace each v that is equal to k+1 by two v's: one equal to k and one equal to 1, and denote the resulting set by $v_1',\ldots,v_{t'}'$. Then $\prod_{i=1}^{t} (1-\alpha_{v_i}') = \prod_{i=1}^{t'} (1-\alpha_{v_i'}) \geq 1-\alpha_0^*$, and this completes the proof.

Having fixed $\alpha_0 = \alpha_s = \alpha_{s-1}$ at α_0^*, how should we choose the remaining α's? In order to have a reasonable chance of detecting existing inhomogeneities irrespective of their pattern we should like to have none of the α's too small. In fact, at the end of the section we reproduce a set of α-values which for $s \leq 21$ will keep α_0 less than 0.1 without letting any of $\alpha_2,\ldots,\alpha_s$ fall below 0.01. These considerations suggest a minimax approach: namely subject to $\alpha_0 \leq \alpha_0^*$ we wish to maximize $\min(\alpha_2,\ldots,\alpha_s)$. In view of Theorem 6.5.5, this is achieved by maximizing α_2.

6.5.6 **Theorem.** The maximum value of α_2 for a multistage procedure which is separating and satisfies (6.5.12) is given by

(6.5.18) $$\alpha_2 = 1-(1-\alpha_0^*)^{[s/2]^{-1}}$$

where, as usual, [A] denotes the largest integer $\leq A$.

Proof. Instead of fixing α_0 at α_0^* and maximizing α_2, it is more convenient to invert the problem and fix α_2 at, say, α and minimize α_0. The theorem will be proved by showing that the resulting minimum value of α_0 is

(6.5.19) $$\alpha_0^* = 1-(1-\alpha)^{[s/2]}.$$

Suppose first that s is even. Since α_2 is fixed at α, it then follows from the fact that the procedure has been assumed to be separating so that the right hand side of (6.5.10) can be made arbitrarily close to α_0^*. This is seen by using the separability of the configuration $v_1 = \ldots = v_{[s/2]} = 2$.

When s is odd, we put an additional v equal to 1 and use separability of the resulting configuration.

In either case it follows that

(6.5.20) $$\sup_{v_1,\ldots,v_t} [1 - \prod_{i=1}^{t}(1-\alpha_{v_i})] = \sup \alpha(P_1,\ldots,P_s) \geq \alpha_0^*.$$

To complete the proof, we must exhibit values of $\alpha_2,\ldots,\alpha_s$ for which the left hand side of (6.5.20) equals to α_0^*. To this end consider the procedure defined by

(6.5.21) $$\alpha_3 = \ldots = \alpha_s = \alpha.$$

The right hand side of (6.5.10) then becomes $1-(1-\alpha)^{t'}$ where
(6.5.22) $\qquad t' =$ number of v's > 1.
This quantity takes on its maximum value for $t' = [s/2]$ and this completes the proof.

There have been suggested several methods for testing the different hypotheses simultaneously. One of the most commonly used is the studentized maximum modulus method (see Miller (1966) or Roy and Bose (1953)) which consists in rejecting all hypotheses having an absolute value of its t statistic greater than a given number determined by the number of hypotheses, the degree of freedom in the t statistics and the level of significance.

The ordinary studentized maximum modulus method can be improved by constructing a corresponding closed test due to Marcus, Peritz and Gabriel (1976), or an equivalent sequentially rejective test of Holm (1977). In the test the absolute values of the t statistics are first ordered. Then the maximal absolute value is compared to the usual studentized maximum modulus limit. If it is above the limit the corresponding hypothesis is rejected and the next absolute value is compared with the studentized maximum modulus limit for one further hypothesis test. The procedure continues in this way as long as new hypotheses can be rejected. This procedure has a higher power than the ordinary studentized maximum modulus procedure and the difference is non-neglectable.

Marcus, Peritz and Gabriel (1976) have pointed out the lack (and need) of two-sided closed procedures, i.e. closed procedures completed with directional statements about the parameter in case of rejection. One possibility of obtaining directional statements is to use the simple and general sequentially rejective Bonferroni test proposed by Holm (1977, 1979) as described below.

When the n hypotheses H_1, H_2,...,H_n are tested separately by using tests with a level $\frac{\alpha}{n}$ it follows immediately from the Bonferroni inequality that the probability of rejecting any true hypotheses is smaller than or equal to α. This constitutes then a multiple test procedure with the multiple level of significance α for free combination - the classical Bonferroni multiple test procedure.

The separate tests in the classical Bonferroni multiple test are usually performed by using some test statistics, which we will denote by $Y_1,...,Y_n$. We suppose now that this is the case, and also that these test statistics have a tendency of obtaining greater values when the corresponding hypothesis is not true. The critical level $\hat{\alpha}_k(y)$ for the outcome y of the test statistic Y_k is then equal to the supremum of the probability $P(Y_k \geq y)$ when the hypothesis H_k is true. Defining now the obtained levels R_1, R_2,...,R_n by

$$R_k = \hat{\alpha}_k(Y_k)$$

the classical Bonferroni test can be performed by comparing all the obtained levels R_1, R_2,...,R_n with $\frac{\alpha}{n}$.

Holm (1979) proposes a sequentially rejective Bonferroni test defined by the obtained levels. Denoting by $R^{(1)}$, $R^{(2)}$,...,$R^{(n)}$ the ordered obtained levels and by $H^{(1)}$, $H^{(2)}$,...,$H^{(n)}$ the corresponding hypotheses. The procedure is defined as follows: If $R^{(1)} > \frac{\alpha}{n}$, we accept $H^{(1)}, H^{(2)},...,H^{(n)}$; otherwise, we reject $H^{(1)}$ and check if $R^{(2)} > \frac{\alpha}{n-1}$ or not, continuing this process until no further rejection can be done. This can happen either by accepting all remaining hypotheses or rejecting the last hypothesis $H^{(n)}$.

For the case of testing simultaneously a number of hypotheses by independent test statistics, a recent result by Shaffer (1978) shows that under quite general conditions the directional statements can be made "free of cost" in the sense that a sequentially rejective test with two-sided alternatives can be completed with directional statements without increasing the risk of making false statements above the prescribed level.

Holm (1980) has proved the same type of results for the sequentially rejective refinement of the studentized maximum modulus test. It will thus be shown that the sequentially rejective refinement of the studentized maximum modulus test described earlier can be completed with directional statements without exceeding the prescribed level of the probability of making false statements.

References

[1] Bechhofer, R. E. (1954). A single-sample multiple decision procedure for ranking means of normal populations with known variances. Ann. Math. Statist. 25, 16-39.

[2] Brownie, C. and Kiefer, J. (1977). The ideas of conditional confidence in the simplest setting. Comm. Statist.-Theor. Meth. A6(8), 691-751.

[3] Duncan, D. B. (1955). Multiple range and multiple F-tests. Biometrics 11, 1-42.

[4] Dunnett, C. W. (1970). Multiple comparisons. Statistics in Endocrinology, Ed. J. W. McArthur and T. Colton, Cambridge: MIT Press.

[5] Einot, I. and Gabriel, K. R. (1975). A study of the powers of several methods of multiple comparisons. JASA 70, 574-583.

[6] Gupta, S. S. (1956). On a decision rule for a problem in ranking means. Inst. Statist. Mimeo. Ser. No. 150, Univ. of North Carolina, Chapel Hill.

[7] Gupta, S. S. and Huang, D. Y. (1975). On subset selection procedures for Poisson populations and some applications to multinomial selection problems. Applied Statistics (Gupta, R. P. Ed.), North-Holland Publishing Co., Amsterdam, 97-109.

[8] Gupta, S. S., Huang, D. Y. and Huang, W. T. (1976). On ranking and selection procedures and tests of homogeneity for binomial populations. Essays in Probability and Statistics (Eds. S. Ikeda et al), Shinko Tsusho Co. Ltd., Tokyo, Japan, 501-533.

[9] Gupta, S. S. and Huang, D. Y. (1977). Some multiple decision problems in analysis of variance. Comm. Statist. A-Theory Methods 6, 1035-1054.

[10] Gupta, S. S. and Nagel, K. (1971). On some contributions to multiple decision theory. Statistical Decision Theory and Related Topics (Gupta, S. S. and Yackel, J.), Academic Press, New York.

[11] Gupta, S. S. and Wong, W. Y. (1976). On subset selection procedures for Poisson processes and some applications to the binomial and multinomial problems. Operations Research XXIII (R. H. Karlsruhe, Ed.), Verlag Anton Hain Meisenhem Am Glan.

[12] Hartley, H. O. (1955). Some recent developments in analysis of variance. Communications in Pure and Applied Mathematics 8, 47-72.

[13] Holm, S. A. (1977). Sequentially rejective multiple test procedures. Report 1977-1, Univ. of Umea, Sweden.

[14] Holm, S. A. (1979). A simple sequentially rejective multiple test procedure. Scand. J. Statist. 6, 65-70.

[15] Holm, S. A. (1980). A stagewise directional test based on T statistics. Report, Chalmers Univ. of Technology, Göteborg, Sweden.

[16] Keuls, M. (1952). The use of the "studentized range" in connection with an analysis of variance. Euphytica 1, 112-122.

[17] Kiefer, J. (1975). Conditional confidence approach in multidecision problems. Proc. 4th Dayton Multivariate Conf. ed. P. R. Krishnaiah, Amsterdam: North Holland Publishing Co., 143-158.

[18] Kiefer, J. (1976). Admissibility of conditional confidence procedures. Ann. Math. Statist. 4, 836-865.

[19] Kiefer, J. (1977). Conditional confidence statements and confidence estimators (With comments). JASA 72, 789-827.

[20] Lehmann, E. L. (1957). A theory of some multiple decision problems I II. Ann. Math. Statist. 28, 1-25, 547-572.

[21] Lehmann, E. L. and Shaffer, J. P. (1977). On a fundamental theorem in multiple comparisons. JASA 72, 576-578.

[22] Lehmann, E. L. and Shaffer, J. P. (1979). Optimum significance levels for multistage comparison procedures. Ann. Statist. 7, 27-45.

[23] Marcus, R., Peritz, E. and Gabriel, K. R. (1976). On closed testing procedures with special reference to ordered analysis of variance. Biometrika 63, 655-660.

[24] Miller, R. G. Jr. (1966). Simultaneous Statistical Inference. New York, McGraw-Hill Book Co.

[25] Miller, R. G. Jr. (1977). Developments in multiple comparisons. JASA 72, 779-788.

[26] Newman, D. (1939). The distribution of the range in samples from the normal population, expressed in terms of an independent estimate of standard deviation. Biometrika 31, 20-30.

[27] O'Neil, R. and Watherill, G. B. (1971). The present state of multiple comparison methods. J. Roy. Statist. Ser. B 33, No. 2, 218, 250.

[28] Roy, S. N. and Bose, R. C. (1953). Simultaneous confidence interval estimation. Ann. Math. Statist. 24, 513-536.

[29] Ryan, T. A. (1959). **Multiple** comparisons in psychological research. Psychological Bulletin 56, 26-47.

[30] Shaffer, J. P. (1978). Control of directional errors with stagewise multiple test procedures. Accepted for publication in the Ann. Statist.

[31] Spjøtvol, E. (1972). On the optimality of some multiple comparison procedures. Ann. Math. Statist. 43, 398-411.

[32] Spjøtvol, E. (1974). Multiple testing in the analysis of variance. Scand J. Statist. 1, No. 3, 97-114.

[33] Tukey, J. W. (1953). The problem of multiple comparisons. Unpublished manuscript, Princeton University.

SUBJECT INDEX

	Page
Associated random variables	19
Best populations	43
Composite hypotheses	83
Conditional inference	80
Decreasing in transposition function	5
Essentially completeness	34
Generalized monotone likelihood ratio	51
Good populations	38
Hypothesis testing	29
Indifference zone selection procedures	31
Just rule	46
Majorization	21
Γ-minimax principle	34
Monotone likelihood ratio	1
Monotone procedure	58
Monotone selection rule	52
Multiple comparison problems	33
Multiple comparison procedure	85
Multiple decision problems	29
Multiple range test	90
Multistage comparison procedures	94
Negatively quadrant dependent	15
Negatively regression dependent	16
Permutation invariant rule	47
Positively quadrant dependent	15
Positive likelihood ratio dependence	18
Positively regression dependent	16
Prior information	34
Property M	2
Restricted minimax approach	34
Schur concave	21
Schur convex	21
Schur-procedures	46
Sequentially rejective Bonferroni test	100
Simultaneous statistical inference	85
Strict monotone likelihood ratio	59
Stochastically increasing property	1,3
Strongly positive quadrant dependent	16

	page
Strongly unimodal	23
Subset selection procedures	31
Total monotone likelihood ratio	5
Total positivity of order 2	1
Total stochastic monotone property	5
Unimodal	9

Lecture Notes in Statistics

Vol. 1: R. A. Fisher: An Appreciation. Edited by S. E. Fienberg and D. V. Hinkley. xi, 208 pages, 1980.

Vol. 2: Mathematical Statistics and Probability Theory. Proceedings 1978. Edited by W. Klonecki, A. Kozek, and J. Rosiński. xxiv, 373 pages, 1980.

Vol. 3: B. D. Spencer, Benefit-Cost Analysis of Data Used to Allocate Funds. viii, 296 pages, 1980.

Vol. 4: E. A. van Doorn: Stochastic Monotonicity and Queueing Applications of Birth-Death Processes. vi, 118 pages, 1981.

Vol. 5: T. Rolski, Stationary Random Processes Associated with Point Processes. vi, 139 pages, 1981.

Vol. 6: S. S. Gupta and D.-Y. Huang, Multiple Statistical Decision Theory: Recent Developments. viii, 104 pages, 1981.

Vol. 7: M. Akahira and K. Takeuchi, Asymptotic Efficiency of Statistical Estimators. viii, 242 pages, 1981.

Vol. 8: The First Pannonian Symposium on Mathematical Statistics. Edited by P. Révész, L. Schmetterer, and V. M. Zolotarev. vi, 308 pages, 1981.

Springer Series in Statistics

L. A. Goodman and W. H. Kruskal, Measures of Association for Cross Classifications. x, 146 pages, 1979.

J. O. Berger, Statistical Decision Theory: Foundations, Concepts, and Methods. xiv, 420 pages, 1980.

R. G. Miller, Jr., Simultaneous Statistical Inference, 2nd edition. 300 pages, 1981.

P. Brémaud, Point Processes and Queues: Martingale Dynamics. 352 pages, 1981.

Lecture Notes in Mathematics

Selected volumes of interest to statisticians and probabilists:

Vol. 532: Théorie Ergodique. Actes des Journées Ergodiques, Rennes 1973/1974. Edited by J.-P. Conze, M. S. Keane. 227 pages, 1976.

Vol. 539: Ecole d'Ete de Probabilités de Saint-Flour V–1975. A. Badrikian, J. F. C. Kingman, J. Kuelbs. Edited by P.-L. Hennequin. 314 pages, 1976.

Vol. 550: Proceedings of the Third Japan-USSR Symposium on Probability Theory. Edited by G. Maruyama, J. V. Prokhorov. 722 pages, 1976.

Vol. 566: Empirical Distributions and Processes. Selected Papers from a Meeting at Oberwolfach, March 28–April 3, 1976. Edited by P. Gänssler, P. Revesz. 146 pages, 1976.

Vol. 581: Séminaire de Probabilités XI. Université de Strasbourg. Edited by C. Dellacherie, P. A. Meyer, and M. Weil. 573 pages, 1977.

Vol. 595: W. Hazod, Stetige Faltungshalbgruppen von Warscheinlichkeitsmassen und erzeugende Distributionen. 157 pages, 1977.

Vol. 598: Ecole d'Ete de Probabilités de Saint-Flour VI–1976. J. Hoffmann-Jorgensen, T. M. Ligett, J. Neveu. Edited by P.-L. Hennequin. 447 pages, 1977.

Vol. 636: Journées de Statistique des Processus Stochastiques, Grenoble 1977. Edited by Didier Dacunha-Castelle, Bernard Van Cutsem. 202 pages, 1978.

Vol. 649: Séminaire de Probabilités XII. Strasboug 1976–1977. Edited by C. Dellacherie, P. A. Meyer, and M. Weil. 805 pages, 1978.

Vol. 656: Probability Theory on Vector Spaces, Proceedings 1977. Edited by A. Weron. 274 pages. 1978.

Vol. 672: R. L. Taylor, Stochastic Convergence of Weighted Sums of Random Elements in Linear Spaces. 216 pages, 1978.

Vol. 675: J. Galambos, S. Kotz, Characterizations of Probability Distributions. 169 pages, 1978.

Vol. 678: Ecole d'Ete de Probabilités de Saint-Flour VII–1977. D. Dacunha-Castelle, H. Heyer, and B. Roynette. Edited by P.-L. Hennequin. 379 pages, 1978.

Vol. 690: W. J. J. Rey, Robust Statistical Methods. 128 pages, 1978.

Vol. 706: Probability Measures on Groups, Proceedings 1978. Edited by H. Heyer. 348 pages, 1979.

Vol. 714: J. Jacod, Calcul Stochastique et Problemes de Martingales. 539 pages, 1979.

Vol. 721: Seminairé de Probabilités XIII. Proceedings, Strasbourg, 1977/78. Edited by C. Dellacherie, P. A. Meyer, and M. Weil. 647 pages, 1979.

Vol. 794: Measure Theory, Oberwolfach 1979, Proceedings, 1979. Edited by D. Kolzöw. 573 pages, 1980.

Vol. 796: C. Constantinescu, Duality in Measure Theory. 197 pages, 1980.

Vol. 821: Statistique non Paramétrique Asymptotique, Proceedings 1979. Edited by J.-P. Raoult. 175 pages, 1980.

Vol. 828: Probability Theory on Vector Spaces II, Proceedings 1979. Edited by A. Weron. 324 pages, 1980.

RAYMOND H. FOGLER LIBRARY
DATE DUE